To
Prof. J. H. C. Coffin
with the regards
of the Author

LECTURE-NOTES

ON

PHYSICS.

By ALFRED M. MAYER, PhD.,

Professor of Physics in the Lehigh University, Bethlehem, Pa.

PART I.—Containing

§ I. Definitions and Introduction to the Inductive Method.

§ II. Instruments used in Precise Measurements.

§ III. Methods of Precision.

§ IV. Manners of Expressing a Law—Law evolved from the Numerical results of Observations and Experiments.

§ V. The General Properties of Matter—The Constitution of Matter according to the Molecular Hypothesis.

§ VI. Capillary Attraction.

PHILADELPHIA:

FROM THE JOURNAL OF THE FRANKLIN INSTITUTE.

1868.

"Who hath measured the waters in the hollow of His hand, and meted out heaven with the span, and comprehended the dust of the earth in a measure, and weighed the mountains in scales, and the hills in a balance?" (ISAIAH xl. 12.)

"Thou hast ordered all things in measure and number and weight." (BOOK OF THE WISDOM OF SOLOMON, chap. xi. 20.)

"It ought to be eternally resolved and settled that the understanding cannot decide otherwise than by induction, and by a legitimate form of it."

"Francis of Verulam thought thus, and such is the method which he determined within himself, and which he thought it concerned the living and posterity to know."

"Let no one enter here who is ignorant of geometry." (PLATO.)

LECTURE-NOTES ON PHYSICS.

§ I. *Definitions and Introduction to the Inductive Method.*

Physical science is the knowledge of the laws (*c*) of the phenomena (*b*) of matter (*a*).

(*a*.) Matter is that which affects the senses. It always presents the three dimensions, or, in other words, occupies space.

It exists throughout all known space as a highly elastic and rare medium called *ether* (proof of this given in optics), and in this medium circulate dense bodies of spheroid forms, separated from each other by distances which are immense when compared with the size of these bodies; such are the planets and asteroids of celestial space, and the earth on which we live.

Shooting-stars or Aërolites are celestial bodies of smaller size, which at certain periods (November 13th and August 10th) cross the orbit of the earth.

Comets are highly rarified nebulous bodies of various aud changing forms, circulating in orbits which are so eccentric, that they are not visible to us in their remote situations; while at other times they are much nearer to the sun than any of the planets. The comet of 1680, when nearest to the sun, was only one-sixth of the sun's diameter from his surface.

The motion of comets retarded by the resistance of the *ether?*

The above is a statement of all known matter, and physical science is that branch of knowledge which considers all the various phenomena which this matter presents.

From the investigations of chemists on terrestrial matter, and from the spectroscopic examination of celestial bodies, all matter can be resolved into (at present) sixty-three *elements.*

According to the atomic theory see (§V.) all matter consists of exceedingly minute, absolutely hard and unchangeable *atoms*, separated from each other by distances which are very great when compared with the dimensions of these atoms.

As in interplanetary space exists the rare ether, so interatomic space is occupied by the same elastic medium. (Proofs of above given under Undulatory Theory of Light and Heat.) Matter exists in the three states of solids, liquids, and gases.

Each organ of our senses is so constructed that it can only take cognizance of those effects which it was specially designed to receive. Thus, through the eye we perceive light, but not sound; and the ear does not take cognizance of light, nor of flavor, nor of odor.

All the senses are modifications of touch.

In touch, taste, and smell the organs of these senses come in contact with the matter that we touch, taste, and smell. In the cases of sound, of light, and of heat, these effects are propagated through an intervening elastic medium whose particles, vibrating in unison with the particles of the sounding, luminous, or heated body, cause the nerves of the ear, of the eye, or of the skin to pulsate; and thus we have special sensations which we severally interpret as sound, light, and heat.

Experiments showing the transmission of sound—vibrations through the air, through a liquid and through a long rod to the ear. The human body can be the intervening vibrating medium to transmit the sound.

Experiment. Sound not propagated through a vacuum. *Ether* is the intervening vibrating medium in the case of light and heat.

A *shadow* is not matter, because it does *not* affect the senses. The light reflected from the surface, contiguous to the shadow, affects the eye; but from where the shadow is, no effect emanates, and the consciousness of the *absence of an effect* on that part of the eye where the shadow is projected gives us the idea of darkness.

According to Helmholtz and Du Bois Raymond an effect is propagated along the nerves with a velocity of 93 feet in one second.

(*b*.) Phenomena. A collection of associated facts.

Example. Compression of air. Here the associated facts are two. (1) The volumes of air corresponding to (2) different pressures. The fall of a stone, associated facts. (1) Spaces fallen through, and (2) the times occupied in falling through the spaces.

Every physical phenomenon is either motion or the result of motion.

Examples. In the phenomena of astronomy, mechanics, acoustics, light, heat, electricity, chemistry, botany, zoology.

The idea of *motion* necessarily embraces two others viz: that of *space* and that of *time;* for a motion must take place in space, and must happen with more or less rapidity, whence our idea of time. (See Instruments used to Measure Time, under § II.)

(*c.*) Law. The expression of the general relation which pervades a class of facts, or the rule according to which a cause acts.

Examples. The law of the compression of gases; volumes of gases are inversely as the pressures to which they are subjected. Gravitation; the gravitating effect is directly as the masses, and inversely as the square of the distance between the gravitating bodies. Sound, light, heat, electric and magnetic attraction diminish in intensity inversely as the squares of the distances from their centres of origin. Reflection and refraction of light.

Universality of these general relations throughout the universe. Importance of these general expressions in assisting the memory. A law embraces in itself all the myriads of facts of which it is the generalization.

The use of a law is shown in its application to the practical purposes of life.

Examples. Application of the law of the compression of gases; of the reflection and refraction of light; of the pressure of steam at different temperatures; of the laws of dynamic electricity.

Without the knowledge of the law of a class of facts, we are obliged to make experiments to solve each individual problem, while if we have the knowledge of the law, the solution of the problem is a mere deduction from the law to which it belongs. Examples drawn from above laws.

The money value of a law to the engineer, and to the industrial and commercial world.

The discovery of laws is the object of science.

The logical arrangement of all the known physical laws constitutes physical science.

"*Classification of the Physical Sciences.*

I. INORGANIC.

a. Celestial Phenomena.

1. Astronomy.
2. Meteorology (part of).

b. Terrestrial Phenomena.

1. Physics.
2. Chemistry.
3. Geology, including Mineralogy.

II. ORGANIC.

1. Botany.
2. Zoology."

PROF. J. HENRY.

Physics considers the laws of the general phenomena of matter, or is the study of the laws of those phenomena which do not bring about a permanent change in the constitution of bodies.

Examples. Fall of a stone, compression of gases, expansion of bodies by heat, reflection and refraction of light, &c.

Chemistry considers the peculiar or individual phenomena of bodies, or is that science which studies those phenomena which bring about a permanent change in the nature of bodies.

Examples. Burning of a candle, rusting of iron, union of sulphur and iron, union of oxygen and hydrogen.

The fundamental principle on which we repose in all our scientific reasoning is, that *the laws of nature are constant*, or, in other words, that *like causes will always produce like effects.*

"The mode of reasoning, in physical science, which is the most generally to be adopted, depends on this axiom which has always been essentially concerned in every improvement of natural philosophy, but which has been more and more employed ever since the revival of letters, under the name of induction, and which has been sufficiently discussed by modern metaphysicians. That like causes produce like effects, or that in similar circumstances similar consequences ensue, is the most general and most important law of nature; it is the foundation of all analogical reasoning, and is collected from constant experience by an indispensable and unavoidable propensity of the human mind." (Dr. Thomas Young's *Lectures on Natural Philosophy*, Lecture I., page 11. London, 1845).

"Most of the phenomena of nature are presented to us as the complex results of the operation of a number of laws."

Examples. In astronomy, in sound, in light, in heat, and in electricity.

"We are said to explain or give the cause of a simple fact when we refer it to the law of the phenomena to which it belongs, or to a more general fact; and a compound one when we analyze it and refer its several parts to their respective laws.

"The indefinite use of the term cause has led to much confusion and error. We distinguish two kinds of causes, *intelligent* and *physical*.

"By an *intelligent cause* is meant the volition of an intelligent and efficient being producing a definite result.

"By a *physical cause*, scientifically speaking, nothing more is unstood than the law to which a phenomenon can be referred.

"Thus we give the physical cause of the fall of a stone or of the elevation of the tides, when we refer these phenomena to the law of gravitation; and the intelligent cause (sometimes also called the moral or efficient cause), when we refer this law to the volition of the Deity."

[It is generally easy to know to what physical cause we should refer a class of physical phenomena; but what is often difficult, is to determine the nature and manner of acting of that cause, and to show how the phenomena and their laws are consequences of its properties. This is the nicest point in physical science.]

"In the investigation of the order of nature, two general methods have been proposed—I. The *à priori* method, and II. The inductive method.

"I. The *à priori method* consists in reasoning downwards from the original cognitions, which, according to the *à priori* philosophy, exist in the mind relative to the nature of things, to the laws and the phenomena of the material universe."

[Examples given of ancient and medieval *à priori* reasoning in their explanations of the motions of the heavenly bodies, and of the law of falling bodies.]

"II. The *inductive method*, which is the inverse of the other, is founded on the principle that all our knowledge of nature must be derived from experience. It therefore commences with the study of phenomena, and ascends from these by what is called the inductive process, to a knowledge of the laws of nature. It is by this method that the great system of modern physical science has been established. It was used in a limited degree by the ancients, and especially by Aristotle, but its importance was never placed in a conspicuous light until the publication of the *Novum Organum* of Bacon in 1620."

"In the application of the *inductive method* to the discovery of the laws of nature, four processes are usually employed.

"1. *Observation*, which consists in the accumulation of facts, by watching the operations of nature as they spontaneously present themselves to our view.

"This is a slow process, but it is almost the only one which can be employed in some branches of science,—for example, in astronomy.

"2. *Experiment*, which is another method of observation, in which we bring about, as it were, a new process of nature by placing matter in some new condition."

[In making an experiment we should not only evolve the phenomena whose laws we wish to determine, but should also so produce these phenomena that we can *measure* their related parts.]

"This is a much more expeditious process than that of simple observation, and has been aptly styled the method of cross-questioning or interrogating nature.

"The term *experience* is often used to denote either observation, or experiment, or both.

["A most important remark, due to Herschel, regards what are called *residual phenomena.* When, in an experiment, all known causes being allowed for, there remain certain unexplained effects (excessively slight it may be), these must be carefully investigated, and every conceivable variation of arrangement of apparatus etc. tried; until, if possible, we manage so to exaggerate the residual phenomenon as to be able to detect its cause. It is here, perhaps, that in the present state of science we may most reasonably look for extensions of our knowledge; at all events, we are warranted by the recent history of Natural Philosophy in so doing. Thus, for example, the slight anomalies observed in the motion of Uranus led Adams and Le Verrier to the discovery of a new planet; and the fact that a magnetized needle comes to rest sooner when vibrating above a copper plate than when the latter is removed, led Arago to what was once called magnetism of rotation, but has since been explained, immensely extended, and applied to most important purposes. In fact, this accidental remark about the oscillation of a needle led to facts from which, in Faraday's hands, was evolved the grand discovery of the Induction of Electrical Currents by magnets or by other currents." *Natural Philosophy*, by Sir William Thomson and P. G. Tait. Oxford, 1867, p. 308.]

"3. *The Inductive Process*, or that by which a general law is inferred from particular facts. This consists generally in making a number of suppositions or guesses as to the nature of the law to be discovered, and adopting the one which agrees with the facts. The law thus adopted is usually further verified by making deductions from it and testing these by experiment. If the result is not what was anticipated, the expression of the law is modified, perhaps many times in succession, until all the inferences from it are found in accordance with the facts of experience."

[Examples of induction in the discovery, by Newton, of the law of gravitation; of the composition of light; of Wells's theory of dew; of Ampére's laws of electro-dynamics.]

"4. *Deduction*, which is the inverse of induction, consists in reasoning downwards from a law, which has been established by induction, to a system of new facts. In this process the strict logic of mathematics is employed, the laws furnished by induction standing in the place of axioms. Thus all the facts relative to the movements of the heavenly bodies have been derived, by mathematical reasoning, from the laws of motion and universal gravitation." (Prof. J. Henry.)

[*Examples* of deduction. The discovery of the planet Neptune by Le Verrier and Adams; of conical refraction by Hamilton and Lloyd. Laplace's work *Méchanique Céleste.*]

"Bacon's greatest merit cannot consist, as we are so often told that it did, in exploding the vicious method pursued by the ancients of flying to the highest generalizations for it, and deducing the middle principles from them; since this is neither a vicious nor an exploded method, but the universally accredited method of modern science, and that to which it owes its greatest triumphs. The error of ancient speculation did not consist in making the largest generalizations first, but in making them without the aid or warrant of rigorous inductive methods, and applying them deductively without the needful use of that important part of the deductive method termed verification." (J. Mills's *System of Logic*, vol. ii., page 524–5.)

The agreement of the deductions which we make from a law, with the facts of subsequent observations and experiments, is the only test of our having truly arrived at a law.

Bacon well remarked, that the test of true science was its utility, and Franklin was well aware of this when a friend, asking him the use of electricity, he replied, characteristically, "What is the use of a baby?"

"When one system of facts is similar to another, and when, therefore, we infer that the law of the one is similar to the law of the other, we are said to reason from *analogy*.

"This kind of reasoning is of constant use in the process of induction, and is founded on our conviction of the uniformity of the laws of nature.

"In the process of the discovery of a law, the supposition which we make as to its nature must be founded on a physical analogy between the facts under investigation and some other facts of which the law is known. One successful induction is thus the key to another."

[*Examples.* Young's discovery of the interference of light; Franklin's of the identity of lightning and electricity.]

"A supposition or guess thus made from analogy as to the nature of the law of a class of facts, is usually called an *hypothesis* and sometimes the *antecedent probability.*

"When an hypothesis of this kind has been extended and verified, or, in other words, when it has become an exact expression of the law of a class of facts, it is then called a *theory*.

[Flourens, in his *Eloge sur Buffon*, says: "An *hypothesis* is the explanation of facts by possible causes; a *theory* is the explanation of facts by real causes."]

"Physical theories are of two kinds, which are sometimes called *pure* and *hypothetical*. The one being simply the expression of a law, which is the result of a wide induction, resting on experiment and observation. Such is the theory of universal gravitation; the theory of sound.

"The other consists of an hypothesis combined with the facts of experience Of this kind is the theory of electricity which attributes a large class of phenomena to the operations of an hypothetical fluid endowed with properties, so imagined as to render the theory an expression of the law of the facts." (Prof. J. Henry.)

A theory is therefore an assemblage of the facts and of the laws, with the consequences which may therefrom be deduced, which belong to one and the same physical cause. Thus we have the theory of gravitation; of light; of heat; of sound, &c.

A theory, to be true and complete, should explain to their minutest details all the phenomena produced by the cause, and that, naturally, without its being necessary to modify the given facts and their laws. Also the numerical results deduced by mathematical calculation should agree with those furnished by direct observation. If, moreover, the logical deductions which we obtain enable us to predict new phenomena which experience afterwards confirms, that theory carries with it the most convincing proof of its reality. The theory of gravitation and the undulatory theory of light afford us beautiful and instructive examples.

In order to establish a theory and follow logically the consequences of the principle from which we have started, we are obliged to compare among themselves quantities which are often bound together by very complicated relations. The unaided attention is not equal to such an effort, and hence we continually call in the aid of *mathematical analysis* to assist us in our investigations in physics.

"Strictly speaking, no theory in the present state of science can be considered as an actual expression of the truth.

"It may, indeed, be the exact expression of the laws of a limited class of facts, but in the advance of science it is liable to be merged in a higher generalization or the expression of a wider law.

"It should be recollected that the laws of nature are contingent truths, or such as might be different from what they are, for anything we know—that they can only be established by induction from the facts of experience—that they admit of no other proof than the *à posteriori* one of the exact agreement of all the deductions from them with the actual phenomena of nature, and that no other cause can be assigned for their existence than the will of the Creator.

"The ultimate tendency of the study of the physical sciences is the improvement of the intellectual, moral, and physical condition of our species. It habituates the mind to the contemplation and discovery of truth. It unfolds the magnificence, the order, and the beauty of the material universe, and affords striking proofs of the beneficence, the wisdom, and power of the Creator. It enables man to control the operations of nature, and to subject them to his use." (Prof. J. Henry.)*

§ II. *Instruments used in Precise Measurements.*

A law of physics, as we have seen, consists in the general relation which exists between the numbers which are the final results of our observations and experiments; and therefore, the first subject to be discussed in the exposition of that process by which we arrive at a law, is the description of those *instruments* which have given such minute precision in the numerical determinations of modern physicists.

The *instruments of precision* used in making the measures required by the physicist may be arranged under the five following divisions:—

1. Instruments used to measure . . *Lengths.*
2. " " " . . *Angles.*
3. " " " . . *Volumes.*
4. " " " . . *Weight.*
5. " " " . . *Time.*

1. *Instruments used to measure Lengths—Dividing Engines.*

Standards of Length. Before commencing a measurement of length, the error of the unit we use should be accurately determined by comparison with a government standard, or with a well authenticated copy.

* In § I. we have made several long quotations from an admirable "Syllabus of a Course of Lectures on Physics," by Prof. Joseph Henry; for it was not in our power to have expressed these important fundamental truths so well, and we do not know of any work in which they are so concisely stated.

Two units of length are used in this country: *the yard* (subdivided into feet, inches, and tenths of inches), and *the mêtre* (subdivided into tenths, hundredths, thousandths, and so on decimally).

The yard, originally derived, about 1120, from the length of King Henry the First's arm, was accurately *fixed* by Captain Kater, in 1818, when he obtained the ratio of the length of the *standard English yard* to the length of a pendulum beating seconds (*i. e.* making 86,400 vibrations in one mean *solar* day) *in vacuo* at the level of the sea, in the latitude of Greenwich. He found that the length of such a pendulum, from the point of suspension to the centre of oscillation, was 39·13929 inches of Bird's standard of 1760, at 62° Fahr.

The actual standard of length of the United States is a brass scale of 82 inches in length, prepared for the U. S. Coast Survey, by Troughton of London; meant to be identical with the English Imperial Standard, and deposited in the Office of Weights and Measures at Washington. The temperature at which it is a standard is 62° Fahr., and the yard measure is the length between the 27th and the 63d inches of the scale. (See "Report on the Construction and Distribution of Weights and Measures," by Dr. A. D. Bache. 1857.)

Two copies of the new British standard, viz: a bronze standard, No. 11, and a malleable iron standard, No. 57, have been presented by the British Government to the United States. A series of careful comparisons, made in 1856, by Mr. Saxton, under the direction of Dr. Bache, of the British bronze standard, No. 11, with the Troughton scale of 82 inches, showed that *the British bronze standard yard is shorter than the American yard by* 0·00087 *inch.* So that, in very exact measures with the yard-unit, it is necessary to state whether the standard is of England or of the United States, since 10,000 American feet = 10,000·5803 English feet.

"*The Mêtre* is a standard bar of platina, made by Lenoir in Paris, which has its normal length at the temperature of zero centigrade, or the freezing point. Its length is intended to make it a *natural* standard, and to represent *the ten millionth part* of the terrestrial arc comprised between the equator and the pole, or of a quarter of the meridian. The length of this arc, given by the measurement, ordered for the purpose by the Assemblée Nationale, of the arc of the meridian between Barcelona, through France to Dunkirk (about 9½ degrees), combined with the measurements previously made in Peru and Lapland, gave for the distance of the equator from the pole 5,130,740 *toises.* The standard toise was made in 1735, in Paris, by

Langlois, under the direction of Godin. It is a bar of iron which has its standard of length at the temperature of 13° Reaumur. It is known as the Toise du Pérou, because it was used by the French academicians Bouguer and La Condamine in their measurement of an arc of the meridian of Peru. The toise of Peru = 2·13145 English standard yards. They found an ellipticity of $\frac{1}{334}$, and the length of the mêtre 443·29596 lines of the toise du Pérou, assumed to be 443·296 lines, or 3 feet 11·296 lines. This last quantity was declared in 1799 to be the length of the *legal mêtre*, and *vrai et définitif*, and is the length of Lenoir's platina standard. Later and more extensive measurements in various parts of the globe, however, seem to indicate that this quantity is somewhat too small. The latest and most exact results we now possess, combined and computed by Bessel, would make the quarter of the meridian 10,000,856 mêtres, and the mêtre = 443·29979 Paris lines. Schmidt's computation would make it 443·29977 lines, and both numbers are confirmed by Airy's results. The legal mêtre is thus, in fact, as Dove remarks, *a legalized part of the toise du Pérou*, and this last remains the *primitive standard*. But it must be added that a natural standard, in the absolute sense of the word, is a utopian one, which ever-changing nature never will give us. The mêtre is, for all practical purposes, what it was intended to be, a natural standard; though it must be confessed that, in practice, the question is not whether and how far a standard is a natural or a conventional one, but how readily and accurately it can be obtained, or recovered when lost." (See A. Guyot's Tables, Meteorological and Physical, p. 111.)

Captain Kater in 1818 determined with great care the value of the *mêtre* at a temperature of 32° Fahr., in inches of Shuckburgh's copy (made by Troughton) of Bird's standard yard at 62° Fahr., and found one mêtre at 32° Fahr.; = 39·37079 inches at 62° Fahr. = 39·36850535 United States standard inches at 62°. Therefore the quadrant of the *French* meridian contains 393,707,900 English standard inches.

It has recently been shown by M. Schubert that the equator of the earth is not a great circle, but an ellipse; having its major axis = 41,854,800 feet, and its minor axis = 41,850,007 feet, giving an ellipticity of $\frac{1}{8880}$th. From this it necessarily follows that different quadrants of the earth will differ, being longer when passing near the major axis than when passing in the neighborhood of the minor axis. Consequently, the mêtre is only the $\frac{1}{10000000}$th part of the quadrant passing through Dunkirk. Sir John Herschel computes 4008 feet for the excess of the true quadrant over that assumed as the basis of the metrical system; which makes the French standard $\frac{1}{208}$th of an inch too short.

Sir John Herschel proposes for the British standard an aliquot part of the *polar axis* of the earth, as a natural unit, and shows that by increasing the existing British standard yard (and with it, of course, its subdivisions) by exactly the $\frac{1}{1000}$th part of its present length, the polar axis will contain 500,500,000 of inches each $\frac{1}{1000}$th longer than the present standard inch. Herschel also shows, that by adopting this new value of a foot, that one cubic foot of distilled water at 62° Fahr. will contain 1000 ounces if we increase the present ounce only the $\frac{1}{18}$th of a grain; and thus there will exist a simple relation between the measures of length and of weight. (See Familiar Lectures on Scientific Subjects, Article, "The Yard, Pendulum, and Mêtre," by Sir John Herschel. A. Strahan: London and New York, 1866.)

In 1866, Congress passed an act directing that 15 grammes of the metric system of weights shall be deemed, and be taken for U. S. postal purposes, as the equivalent of one-half ounce avoirdupois.

The new U. S. five-cent piece is made 2 centimetres in diameter and weighs 5 grammes.

See plate I. for the French measure of one decimetre, divided into centimetres and millimetres; and the English measure of four inches divided into halves and tenths.

(The standards of one yard and of one mêtre exhibited.)

Measurement of lines of considerable length. There are two modes of directly measuring lines of considerable length. The first consists in placing in *contact* the ends of rods having an exact standard length, and thus applying these units to the line whose length is required. The second consists in bringing the beginning of one unit to coincide with the end of another, by *bisecting* a fine × mark near the end of one of the rods by the cross-threads in a microscope attached to the end of the other rod. The distance on a rod between the cross-hairs in the microscope at one end and the × mark at the other being equal to a standard length.

With the precise apparatus of the U. S. Coast Survey, a base of ten miles can be measured so accurately that the error made is only about ± $\frac{1}{10}$th of an inch. (See U. S. Coast Survey Report for 1854, p. 103, *et seq.*, for a description of this apparatus for measuring base-lines.)

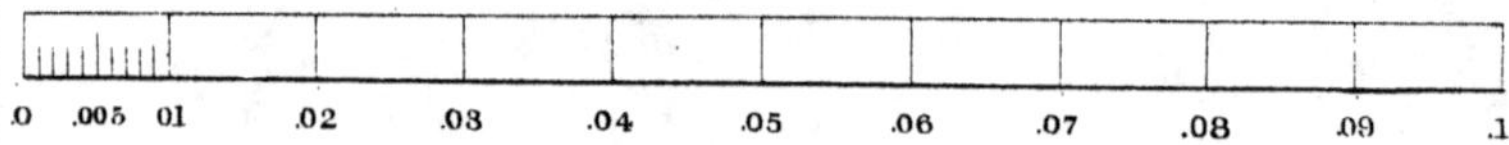

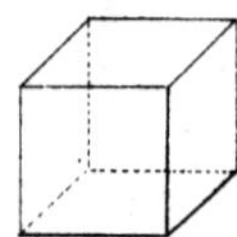

One Cubic Centimêtre of Distilled Water at 4° C.=1 Gramme.

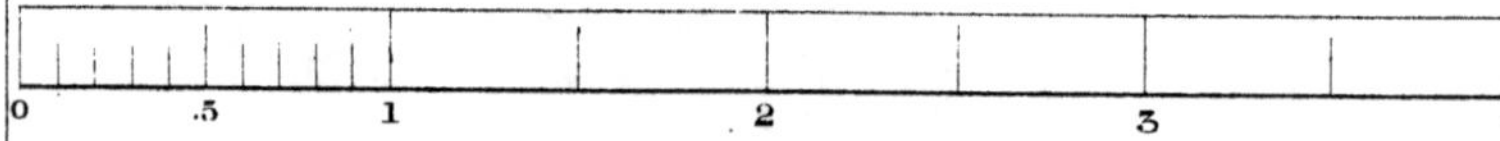

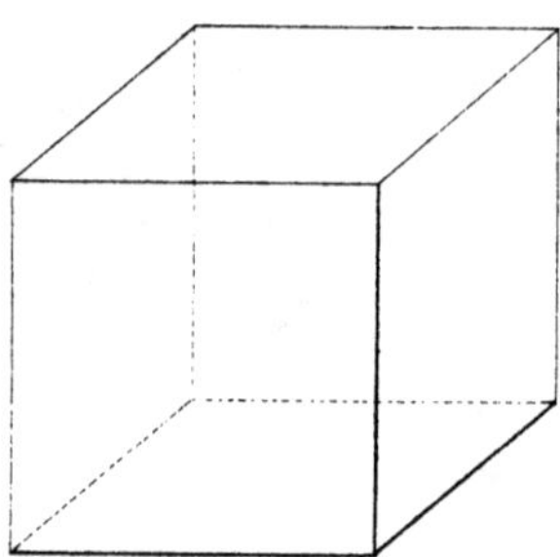

One Cubic Inch of Water at 62° F=252.456 Grains.

The Vernier is a device for subdividing still further the lowest divisions of a scale without dividing directly those divisions by equidistant lines. It consists of a *sliding* scale called the *vernier*, which glides along the length of the main scale, and has on it divisions which *differ* from those of the main scale by a known fraction of the smallest subdivision on the main scale. Thus, suppose we have a main scale of inches divided into tenths, and we wish to divide the tenths of inches into ten parts by a vernier, thus giving us the hundredths of inches: we take a length of *nine-tenths* from the main scale, and place it on the vernier, and we divide this length into *ten* equal parts, which necessarily gives divisions on the vernier, each of which is $\frac{1}{100}$th of an inch less than the smallest division, of tenths, on the main scale; therefore, by seeing where a line of the vernier corresponds with a line on the main scale, the next following lines of the two scales are distant from each other $\frac{1}{100}$th of an inch; the next following lines differ $\frac{2}{100}$ths, and so on; and thus we can subdivide any $\frac{1}{10}$th of an inch of the main scale into hundredths of an inch.

In general, to read to the nth part of a scale division, n divisions of the vernier must equal $n + 1$ or $n - 1$ divisions on the main scale, according as these run in opposite or in similar directions.

Large models exhibited of vernier applied to straight line and to arcs of circles.

Rule for finding the smallest reading by means of the vernier. *Divide the value of the smallest divisions on the main scale by the number of divisions on the vernier.*

This beautiful and very useful invention is due to Pierre Vernier, of France, who first described it in a work entitled "La construction, l' usage et les propriétés du cadran nouveau." Bruxelles, 1631.

The Cathetometer (from Gr. κάθετος vertical height, and μέτρον a measure), consists of a vertical rod, which rotates round its axis, and carries a telescope, provided with a spirit-level, whose line of collimation is at right-angles to the axis of the rod. The rod rests on a base furnished with levelling screws and spirit-levels, so that it can be brought into a truly vertical position. The telescope is attached to a vernier plate, which slides along the length of the rod which is divided into fractions of the mêtre or of the foot. The instrument is used to measure the vertical distance between two points, whether on the same vertical line or not. It is of constant use in making all measures of vertical distances.

(Instrument exhibited, and method of using it shown.)

This instrument was invented by Dulong and Petit, and subsequently improved by Pouillet.

Micrometer Screw (from Gr. μικρος small, and μέτρον a measure) consists of a screw with a large circular head, whose circumference is divided into a certain number of equal parts. Suppose the screw has 50 threads to the inch, and that its circular plate, which rotates with it, is divided into 200 parts; then if the screw is turned a whole revolution, it will advance, in the block in which it turns, $\frac{1}{50}$th of an inch; but if the head is revolved only through $\frac{1}{200}$th of a revolution, the screw will advance the $\frac{1}{10000}$th of an inch. With a good micrometer screw we can measure accurately the $\frac{1}{1000}$th of a millimetre, or about the $\frac{4}{100000}$th of an inch.

(Large model exhibited, also the instrument itself in various forms.)

Spherometer. As its name indicates, it is used to measure spheres; or, more concisely, to determine the radius of any spherical surface; as, for instance, the radius of the surface of a glass lens. It was invented by the optician de Laroue, for the latter determination. It consists in a micrometer screw, supported vertically by a tripod. The points of the feet of the tripod are equally distant from each other; and they, as well as the screw, terminate in fine nicely rounded extremities. The point of the screw and the points of the tripod are brought into the same plane by placing the instrument on the truly plane surface of a slightly roughened glass plate, and bringing the screw's point to bear on the plane with the same degree of pressure as the points of the tripod. This is attained by means of a jointed lever, which moves with the screw, and its shorter arm bearing on the glass, its longer arm is always brought into the same position. (This can only be clearly understood by examining the instrument.) The instrument is now removed to the lens, the radius of whose curvature we desire, and the point of the screw brought down upon the glass till it bears with the same degree of pressure as in the previous instance. The difference of readings on the head of the micrometer screw gives the perpendicular distance of the point of the screw above the plane passing through the points of the feet of the tripod; or, in other words, the height of a segment of a sphere, having for radius the radius of curvature of the lens, and for base the circumscribed circle of the equilateral triangle made by the points of the three feet of the spherometer.

Calling this height h, the radius of the circle passing through the feet of the tripod r, and the radius of the spherical surface of the lens R, we have

$$2\,R - h : r :: r : h; \quad \text{whence}$$
$$2\,R\,h - h^2 = r^2; \quad \text{therefore}$$
$$2\,R = \frac{r^2 + h^2}{h},$$

half the value of 2 R being of course the radius of curvature of the lens.

The spherometer will give accurate measures to the $\frac{1}{1000}$th of a millimetre, and is so minute in its measurements, that if the finger be momentarily placed on the plate of glass under the point of the screw, and we again bring down the screw to this portion of the glass, we find that the heat from the finger has swelled the glass at the part touched into a protuberance!

It is easy to see how this instrument is also used for measuring the thickness of plates.

Comparator. Model of instrument exhibited and described. Invented by Lenoir in 1800, to compare standards of length.

In this instrument, as well as in the following one, we measure a multiple of the quantity we wish to estimate.

Saxton's Reflecting Comparator and Pyrometer. Model of this instrument exhibited, and mode of using it shown.

It is so delicate that it will measure with precision the $\frac{1}{100000}$th of an inch.

"This very ingenious instrument has been applied to comparing the yard of the kind called *end-measure*, in which the distance between the two ends of the bar is the length of the standard yard. One end of the bar to be compared abuts against a fixed support, the other end is free to move, the whole bar being supported horizontally on rollers. The free end presses against a small horizontal bar or slide, connected by a chain, with a vertical axis, carrying a small mirror. As the free end of the bar moves, the slide moves also, and turns the mirror. A scale placed horizontally at any convenient distance from the bar and in the same general direction, is reflected in the mirror, the image being viewed by a telescope placed at the same distance as the scale, and directly over it. The distance of the reflected image of the scale being twice that of the scale from the mirror, the motion of the mirror is shown as on a divided circle of that radius, and the smallest movement of the mirror is measurable."

(See "Report by Dr. A. D. Bache, to the Secretary of the Treasury, on the Construction and Distribution of Weights and Measures," Washington, 1857, p. 15, *et seq.*)

The illustrious Gauss, of Göttingen, in 1827, first used this method of reflection in his measures of the forces of terrestrial magnetism.

It also can be applied to the galvanometer, by attaching its needles to a very light silvered mirror; and thus the smallest motions of the needles can be shown to the largest audience.

Dividing Engines. To divide the straight line into fractional parts, and the circumference of the circle into degrees, minutes, and seconds, machines are used which, when made with care and used with skill, are capable of working with great precision.

To divide the straight line. The simplest apparatus for dividing the straight line is that described by Bunsen in his "Gasometry" (London, 1857). The rod, tube, &c., to be divided is placed in a line with a standard scale, and with long, rigid beam-dividers, having one of its points on a division of the scale we describe, with the other point a line on the rod &c. to be divided; we then move the point of the dividers to the next division of the scale, describe with the other point another line on the rod, and so on.

If the weight of the dividers is partly supported by a cord, passing over a pulley overhead, and attached to a weight, and an assistant, with a glass, carefully places the point on a division of the scale while the operator describes the line, this method will give an accurate copy of the scale from which we transfer the divisions.

The most accurate machine for dividing the straight line consists of a plate placed on Λ shaped guides, and moved in the direction of its length by a screw working in a nut attached to its lower side. The rod &c. to be divided is clamped to the upper surface of the plate, and a cutting tool, which can only move in one and the same plane perpendicular to the length of the plate, cuts the divisions on the rod after each rotation of the screw. If the screw contains ten threads to the centimetre, then at each whole revolution of the screw the plate will advance one millimetre; and any fraction of a millimetre can be cut by rotating the screw that same fraction of a whole revolution.

So minute is the accuracy of which this machine is susceptible, that it is able to cut 75,000 equidistant and parallel lines in the breadth of an inch, and each division is distinctly visible under the highest power of the microscope. A still greater number, probably over 100,000 lines, can be cut in the length of an inch, but they cannot be discerned even by the highest powers of the best microscopes, furnished with the most perfect means of illumination. In this case our mechanical exceeds our optical power.

A graduated series of these lines, increasing in the number to the inch, are used under the name of Nobert's test lines, to determine the defining power of the objectives of microscopes.

Another example of fine engine work is given in "Barton's buttons," which are gold buttons stamped with a hard steel die on which Barton cut hexagonal groups of fine equidistant lines with a diamond. On account of the interference produced in the light reflected from the surface of these buttons, they flash with the brilliancy and the colors of gems.

From the above description of the performance of the dividing engine, it is easy to see how, having fine lines cut at known intervals apart on plates of glass, and placed in the ocular or on the stage of a microscope, we have the means of measuring distances which surpass in minuteness the determinations of the micrometer-screw.

To divide the circumference of the circle. Two methods are used in the graduation of the circumferences of the circles of astronomical and other instruments.

In the first, and probably the most accurate method, the circle to be divided is centered and clamped upon a large *accurately* graduated circular plate, which revolves on a vertical axis, and under a cutting tool which moves only in one and the same vertical plane passing through the centre of the circle. The divisions on the periphery of the large and *originally divided* circle are successively bisected by the cross-wires in the ocular of a fixed microscope, by revolving the plate under the microscope by means of a tangent screw; and at each bisection the tool cuts a line in the border of the circle to be graduated.

In the other method a tangent screw works in the notched edge of a circular plate; and by one whole revolution of the screw the plate is carried round 10′. After each revolution the tool cuts the division. Thus the circle is divided into arcs of 10′ and into smaller fractions of a degree if desired.

The first method is used in Germany; the second, invented by Ramsden, and subsequently improved by Troughton & Simms, and by Gambey, of Paris, is practiced in England and in this country. At the U. S. Coast Survey Office in Washington, can be seen a very superior dividing engine, made by Troughton & Simms, of London.

In both methods the circle of the engine has originally to be divided by the principle of bisection. To bisect a straight line, one point of a beam-compass is placed at one extremity of the line, while with the other point we strike a fine line nearly bisecting the line we wish to divide. From the other extremity of the line an arc is struck with the same radius, and the minute space included between the lines thus formed is bisected with a fine needle-point under a microscope. A scale of equal parts is thus formed with the most extreme care; and having attached to it an accurate vernier, we take in the beam-compass from this scale a length which exactly equals the chord of an arc of 85° 20′ of the circle to be graduated, and with this chord we strike on the circle an arc of 85° 20′. By continued bisections this arc is subdivided into equal parts of 5′ each. The 60° division can be obtained by a chord equal to the radius of the circle, and this arc of 60° bisected gives 30°, which, added to previous arc of 60°, gives an arc of 90°.

The reader is referred to an article on "Graduation," by the celebrated artist Troughton, in the *Edinburgh Encyclopædia*, for the best account extant of this, the most delicate and difficult problem of mechanical art; which has tested to the utmost the combined skill of the first mechanicians, and of the ablest astronomers of modern times.

2. *Instruments used in the Measurement of Angles.*

The circumference of the circle is divided into 360 parts, one of which is called *a degree*—the unit of angle measures. The degree is divided into 60 *minutes*; the minutes into 60 *seconds*; the subdivisions of the second are decimal. The sign of the degree is (°); of the minute (′); of the second (″).

Theodolite, Meridian Circle, Sextant, &c., described from the instruments, and from diagrams.

The Micrometer consists of two spider-lines, stretched parallel to each other, on two frames, each moved by a micrometer screw, so that the threads can be made to approach to, or recede from, each other. The spider-lines are exactly in the focus of the object glass of an achromatic telescope, so that the image of any object formed by the object-glass, and the lines, are in the same plane, and therefore both equally distinct, when viewed through the ocular of the telescope. The value of one turn of the micrometer screw may be estimated in angle units, by causing the spider-lines to embrace the vertical diameter of the sun, when it is on or near the meridian; and then, by obtaining its exact diameter from the Nautical Almanac, we can, after allowance made on our measure for refraction, reduce a turn of the screw to its value in angle.

The invention of the micrometer is due to William Gascoigne, 1640, who also first applied the telescope (invented in 1609, by Galileo) to a graduated circle. Gascoigne was killed, at the early age of 23, while fighting for Charles I., at Marston-Moor, 2d July, 1644.*

The Reading Microscope is a similar contrivance to the Micrometer; the difference being, that the micrometer spider-lines, which cross (×) each other, and are moved by one screw, are placed in the focus of a microscope, instead of in the focus of a telescope. The Reading Microscope is placed over the divisions of a graduated circle, and is so adjusted that five turns of the screw will carry the point of bisection of the cross (×) lines, the length of the smallest division on the circle, which generally equals 5′ of arc. The head of the screw of the reading microscope is divided into 60 parts, so that when the screw is revolved through one of these parts, the cross (×) lines are carried over the circle, a distance equal to *one second* of arc. In those circles used in the most accurate determinations of astronomy, the microscopes read to $\frac{1}{10}$ of a second of arc; and this minute accuracy is absolutely required in many problems now in process of solution by Astronomers—(*e. g.* the parallax of the fixed stars).

The difficulty of measuring with accuracy to a fraction of a second, will be readily appreciated, when we know that one second of arc occupies, on a circle of 6 feet in diameter, a length of only ·0001745 of an inch.

For an excellent method of stretching spider-lines in the micrometer, devised by Lieut. M. F. Maury, see Washington Astronomical Observations, 1845.

Reflecting Goniometer. Used in the measurement of the angles of crystals. Invented by Dr. Wollaston, in 1809. "This instrument will give the inclination of planes of crystals, whose area is less than $\frac{1}{10000}$th part of a square inch, to less than one minute of a degree." (Ency. Metrop. Art. Crystallography.)

* See *Letters of Scientific Men of the Seventeenth Century*, Oxford, 1841, vol. i., p. 33, *et seq.* Letters XIX. and XX., Will. Gascoigne to Oughtred, Dec. 2, 1640.

Instrument exhibited, and method of using explained, by measuring the angle of a crystal.

The Spirit Level is formed of a *slightly* curved glass tube, nearly filled with ether. A bubble of air occupies the highest portion of the convex side of the tube.

Spirit levels have been made of such extreme delicacy, that an inclination of one second of arc, in the plane on which the level rests, would cause the bubble to move three millimêtres; the curvature of the tube of the level, in this case, was 619 mêtres. M. Biot used, in his measurement of the arc of the meridian passing through Dunkirk to Formentera, a level whose bubble moved one millimêtre for an inclination of 1″·79.

By the aid of the levels of the best mechanicians we can safely estimate *tenths of seconds* of arc.

The level appears to have been known to the ancient astronomers of India.

3. *Instruments used in the Measurements of Volumes.*

Volumes of liquids and of gases are measured in graduated cylindrical tubes.

A tube or vessel of any form may be graduated into cubic centimêtres, by marking the levels reached by successive equal portions of mercury poured into the tube or vessel; each portion being equal to the exact weight of one cubic centimêtre of mercury, of the temperature at which the mercury is when the weighing is performed. It is better, however, first to divide the tube into equal divisions of length, say millimêtres, on a dividing engine, and then, with a short tube, whose aperture is closed with a ground glass plate, and whose capacity is exactly one cubic centimêtre, to measure off and pour into the tube, cubic centimêtres of mercury. After each addition, the number of millimêtres to which the mercury rises is noted in a table, and thence the value in capacity of each division of the tube is deduced.

For further information on this subject, see Bunsen's *Gasometry*, London, 1857; and the papers of Regnault, published in the Transactions of the Institute of France.

Volumes of solids, as well as of liquids, may be determined by the estimation of their weight, at a known temperature, and under a known atmospheric pressure. In this method we have also to know the weight of a cubic inch, or of a cubic centimêtre of the solid or liquid. The advantage of the French gramme weight is shown in this determination of volumes, by weight; for the *specific gravity* of a body is equal to the weight of one cubic centimêtre of the substance, in grammes; for one cubic centimêtre of water, at 4° C., weighs one gramme.

4. *Instruments used to Measure Weights.*

Two units of weight are used in this country: *the grain*, fixed by the weight of one cubic inch of distilled water, at 62° F. and 30 inches of the barometer, one of such cubic inches of water weighing 252·456 grains. The other is the French unit, the gramme, which is derived from the weight, in vacuo, of one cubic centimêtre of distilled water, at 4° C. See Plate 1.

The Balance. Saxton's U. S. Standard Balance exhibited and described. Becker's Analytical Chemist's Balance. This balance will weigh to the $\frac{1}{10}$th of a milligramme, with each pan carrying a weight of 50 grammes; or, in other words, its beam will turn with $\frac{1}{500000}$th of that weight, the index being deflected $\frac{1}{3}$d of a division of the scale from the position of equilibrium.

See a paper by W. Crookes, in the *Chemical News* for 19th April, 1867, *On the Correct Adjustment of Chemical Weights;* and the investigations of Prof. W. H. Miller, of Cambridge, in the *R. S. Philosophical Transactions* for 1856.

Method of Double Weighing. Invented by Borda of Paris. Consists in counterpoising a mass, whose weight we desire, by shot, sand, thin copper foil, or pieces of fine wire, and then replacing the mass by weights, until equilibrium is again established. The weights are thus placed in exactly the same conditions as the mass, and therefore this process eliminates all errors arising from unequal length of the arms of the balance, etc. This process is far more accurate than single weighing.

The weights of very small portions of matter, such, for example, as *chemical traces*, can be estimated by the deflection of a very fine filament of glass, one end of which is cemented to the edge of a block, and the other end has hanging to it, a still finer filament, supporting a disk, about $\frac{1}{2000}$th inch thick, made of elder pith, on which is placed the substance, whose weight we determine by the amount of deflection it causes in the glass filament. With this apparatus (the invention of the author), can be estimated the deflection produced by the $\frac{1}{1000}$th of a millimêtre.

See Silliman's *Journal of Science*, vol. xxv., page 39, 1858, for a description of this instrument.

5. *Instruments used in the Measurement of Time.*

Definition of time, from Laplace's Système du Monde, ch. III :—"Time is for us the impression which is left in the memory by a series of events, whose existence we are certain has been successive. Motion can serve to measure it; for a body not being able to be in several places at the same time, can only go from one point to another, by passing successively through all the intermediate points. If at each point of the line which it describes, it is animated with the same force, its motion is uniform, and the parts of that line can measure the time employed to run over them. When a pendulum at the end of each oscillation is found in exactly the same circumstances, the durations of the oscillations are the same, and time can be measured by their number. We can also employ for that measure, the revolutions of the celestial sphere, which are always equal in duration; but all have unanimously agreed to use for that object the motion of the sun, whose returns to the meridian, and to the same equinox, or to the same solstice, form the days and the years."

The duration of one revolution of the celestial sphere constitutes *a sidereal day;* while the time of one mean revolution of the sun equals *a mean solar day.* The sidereal day, which is used by astronomers, is 3m. 56·5s. less than the solar day.

Both days are divided into 24 hours; each hour into 60 minutes; and each minute into 60 seconds; the seconds are sub-divided decimally. The hours are designated by the sign *h;* the minutes by *m;* the seconds by *s*. Both days are divided into 86,400 seconds.

The real unit of time must be the *sidereal day*, being the duration of one revolution of the earth on its axis. From recorded ancient eclipses, it has been computed that the time of this revolution has not altered by $\frac{1}{4300000000}$th of its length, from B. C., 720. From the sidereal day is readily deduced the value of the mean solar day.

The Astronomical Clock A clock consists of a train of wheel-work (moved by the descent of a weight) which maintains and indicates the vibrations of a seconds pendulum. The Astronomical clock is made with every refinement of workmanship; it has generally a "*dead-beat*" escapement, and has always a pendulum which is so compensated that a change of temperature does not alter the length from the point of suspension to the centre of oscillation of the pendulum.

The Chronometer is a large watch, very accurately constructed, having a "*chronometer escapement*" and a carefully compensated balance-wheel. It beats half-seconds.

It is not required that a clock or chronometer should keep the exact time, but it *is* required that the daily *rate* of *variation* from the true time should be constant.

The pendulum clock was invented by Huyghens, of Holland, in 1658; and in the same year Hooke of England applied the spring-balance to time-keepers.

Seconds Stop-Watch exhibited and described.

A Uniformly Revolving Cylinder was first proposed by Dr. Thomas Young, for the measurement of intervals of time. With this apparatus we can, in certain cases, measure the $\frac{1}{10000}$th of a second. "By means of this instrument we may measure, without difficulty, the frequency of the vibrations of sounding bodies, by connecting with them a point, which will describe an undulated path on the revolving cylinder. These vibrations may also serve, in a very simple manner, for the measurement of the minutest intervals of time; for if a body, of which the vibrations are of a certain degree of frequency, be caused to vibrate during the revolution of an axis, and to mark its vibrations on a roller, the traces will serve as a correct index of the time occupied by any part of a revolution; and the motion of any other body may be very accurately compared with the number of alternations marked in the same time, by the vibrating body." From Dr. Young's *Lectures on Natural Philosophy and the Mechanical Arts* (a work to be thoroughly studied by all students of physics), 2d Edit., London, 1845, page 147. See, also, an article on "Apparatus and Experiments," devised by the author, in the Franklin Institute Journal, for November, 1867.

The revolving cylinder has recently been applied, in connection with the electric clock, to the registration of transit observations. The apparatus described, with diagrams.

Kater's Method of determining small intervals of time, by weighing the mercury which flows from a small orifice, under a constant pressure.

Revolving Mirror. Invented by Prof. J. Wheatstone, of London, who, in 1834, measured with it the velocity of electricity. Improved by Arago, and proposed by him to the Institute of France, in December, 1838, to determine the velocity of light. In 1850, Foucault of Paris, so perfected its construction and arrangement, that he measured with it, accurately, the $\frac{1}{15000000}$th of one second of time.

The principle of the apparatus can be rendered clear in a few words. Suppose a ray of light, entering a dark room by a fine slit, falls upon a rapidly revolving mirror. This ray is reflected from the *revolving* mirror to a *fixed mirror*, distant from it say 20 feet. The ray reflected from the fixed mirror, falls again upon the revolving mirror. Now, suppose that in the time the light takes to go from the revolving mirror to the fixed mirror, and to return to the revolving mirror, that by its rotation, the face of the revolving mirror changes its position: then this causes a deviation in the ray reflected on its return to the revolving mirror, compared with its direction when it left the revolving mirror to go to the fixed mirror. Therefore, knowing the time it takes for the mirror to make one revolution, we can compute, from the deviation produced, the time required by the light to go over double the distance from the revolving mirror to the fixed mirror. This distance, in Foucault's experiment of 1850, was only 4 metres (13·12 feet); and the mirror, making 800 turns in a second, produced a deviation in the return ray of $\frac{3}{10}$ millimêtre.

In 1862 Foucault repeated his measurements with a far more precise instrument, and increased the distance of the mirrors from 4 to 20 metres (65 feet 7·4 inches). The result of his measures gives a velocity of light equal to 185,177 miles per second, differing slightly from the measure obtained by astronomic observations, when we use for their reduction the solar parallax, with the increase in amount which it appears from recent observation we should adopt.

The most important result obtained with this apparatus, was the fact that light moves slower in water than in air, the velocities in the two media being as 3 is to 4. In short, the index of refraction expresses the ratio existing between the velocities of light in the two media.

"An admirable experiment in physics, in which, by the power of intellect and manual skill, we have succeeded not only in rendering sensible, but even measurable, the time employed by light to run over a path of 20 metres, although this time barely equals the

$\frac{1}{1500000}$th of a second! and which, if we repeat so as to vary its elements, and thus make evident the constant causes of error which affect the result, appears capable of giving a determination of the velocity of light altogether as precise as that which is deduced from astronomical phenomena!" For further information of this method, see "Essay on the Velocity of Light," by M. Delaunay of the Institute of France; translated for the Smithsonian Institution by Prof. Alfred M. Mayer, 1864.

Calculating Engines.

The great importance of such machines in giving us *faultless* logarithmic, astronomical, nautical, and other tables. Impossible to calculate and print any extensive table without making errors.

For a description of the Arithmetical Machine of Pascal, see *Oevres de Pascal*, vol. ii., page 368, et. seq. Hachette, Paris, 1860.

For information in reference to Charles Babbage's celebrated engines, see *Edinburgh Review*, 1834; Taylor's *Scientific Memoirs*, vol. iii., 1843, London, in which is a description by General Menabrea, of Babbage's Difference Engine, and of his more recent and far superior Analytical Engine, translated, with extensive notes, by the accomplished Lady Lovelace, the only daughter of Lord Byron.

See also, *Passages from the Life of a Philosopher*, by Charles Babbage, London, 1867.

Exhibit the "Tables calculated, stereomoulded, and printed by machinery, London, Longmans & Co." 1857. The engine by which these tables were calculated was made by George and Edward Scheutz, of Stockholm, and is now the property of the Dudley Observatory, at Albany, N. Y.

TABLES FOR THE COMPARISON OF THE FRENCH AND ENGLISH SYSTEMS OF MEASURES AND WEIGHTS.

Measures of Length.

Denomination and Value.		Equivalent in English Standard.
Myriametre	10,000 metres	6·2137 miles.
Kilometre	1,000 "	0·62137 mile = 3,280 feet 10 inches.
Hectometre	100 "	328 feet 1 inch.
Decametre	10 "	393·707 inches.
METRE	1 metre	39·37079 "
Decimetre	$\frac{1}{10}$ "	3·937 "
Centimetre	$\frac{1}{100}$ "	0·3937 "
Millimetre	$\frac{1}{1000}$ "	·03937 "

Measures of Surface.

Denomination and Value.		Equivalent in English Measure.
Hectare	10,000 square metres	2·471 acres.
ARE	100 " "	119·6 square yards.
Centare	1 " metre	1,550 " inches.

Measures of Capacity.

Denomination and Value.		Equivalent in English (Wine) Measure.	
Kilolitre or Stere	1000 litres	= 1 cubic metre	264·17 gallons.
Hectolitre	100 "	= $\frac{1}{10}$ " "	26·417 "
Decalitre	10 "	= 10 " decimetres	2·6417 "
LITRE	1 litre	= 1 cubic decimetre	1·0567 quart.
Decilitre	$\frac{1}{10}$ "	= $\frac{1}{10}$ " "	0·845 gill.
Centilitre	$\frac{1}{100}$ "	= 10 cub. centim.	0·338 fluid oz.
Millilitre	$\frac{1}{1000}$ "	= 1 " "	0·27 " drm.

Weights.

Denomination and Value.		Equivalents in English Avd. Weight.	
Millier or tonneau	= 1,000,000 grms.	= 1 cub. met. of W. @ 4° C.	= 2204·6 lbs.
Quintal	= 100,000 "	= 1 hectolitre " " "	= 220 46 "
Myriagramme	= 10,000 "	= 10 litres " " " "	= 22 046 "
Kilogramme or kilo	= 1000 "	= 1 litre " " " "	= 2·2046 "
Hectogramme	= 100 "	= 1 decilitre " " "	= 3·5274 oz.
Decagramme	= 10 "	= 10 cubic centim. " "	= 0·3527 "
GRAMME	= 1 "	= 1 " " " "	= 15·43235 gs
Decigramme	= $\frac{1}{10}$ "	= $\frac{1}{10}$ " " " "	= 0·5432 "
Centigramme	= $\frac{1}{100}$ "	= 10 cub. millim. " "	= 0·1543 "
Milligramme	= $\frac{1}{1000}$ "	= 1 " " " "	= 0 0154 "

Inch	=	25·39954	millimetres.
Foot	=	3·047945	decimetres.
Mile	=	1609·315	metres.
Square Inch	=	6·451367	square centimetres.
" Foot	=	9·28997	" decimetres.
" Yard	=	83 60971	" "
" Acre	=	·4046711	of a hectare.
" Mile	=	258·9895	hectares.
Cubic Inch	=	16·38618	cubic centimetres.
" Foot	=	28·315312	" decimetres, or litres.
Gallon	=	4·54346	litres.
"	=	277·274	cubic inches.
Grain	=	64·79896	milligrammes.
Pound Avd.	=	453·5927	grammes.

§ III. *Methods of Precision.*

In § II we described various instruments used in precise measurements, and we will now explain certain methods used, both in making a series of measures on a quantity and in subsequently combining them, so that the final result has the greatest possible accuracy. These methods are designated as *methods of precision.* They can be classed under three heads:—

1. Method of means, or of averages.
2. Method of multiplication.
3. Method of successive corrections.

1. *Method of Means.*—If a series of measurements be taken with an instrument of precision on one and the same quantity, it will be found, when all the measures are made in the same circumstances and with equal care, that they will differ from each other by a *small* amount.

The mean value of all the measures is taken when the probability is even that each measure is more or less than the true measure by *a small quantity.* The mean result in this case is the same as if we added the mean to itself as often as there are separate measures, and divided by the number of measures.

The determination of means is of such constant occurrence in physics, astronomy and chemistry, that the discussion of their *degree of precision* is very important. It is founded on the principle of the theory of probabilities; and we will here give a few of its more important results.

Omitting from consideration gross errors, which can always be avoided by operating with care, the causes of the errors of measurements can be classified in two groups, 1. Constant or regular causes; 2. Irregular or accidental causes; whence we have two kind of errors, 1. *Constant or regular errors*, which are reproduced when we repeat the observations in the same circumstances; and 2. *Irregular or accidental errors* of which we are not able to get entirely rid; but which, not being subject to any law connecting them with the circumstances of the measurements, occur indifferently to increase or to diminish the true measure.

Examples. A constant cause of error would be an error in the length of the unit used in measuring the base-line of a trigonometric survey. This error will be *regular* and *constant*, and being made every time the unit is applied to the length to be measured, it will be in proportion to this length. It is evident that this error can be allowed for, when we know its amount, and thus our result is the same as if we operated with an accurate unit. As examples of irregular sources of error, we may instance the measurement or sighting of angles in a survey; and the bisection of a line by the reading microscope, or by the telescope of a catheometer.

It is evident that *constant errors* are not eliminated by increasing the number of observations; but the *accidental errors* tend to disappear from the mean as we obtain it from a greater number of measures; for this class of errors are as likely to be in one direction as in another in successive trials, and therefore the mean of an *infinite* number of measurements on the same quantity would give the greatest attainable precision, for the quantities would then exactly balance each other in excess and defect.

To find the degree of precision of a mean we divide the whole series of observations into two or three groups, selected at random, and we calculate for each its mean. If these means differ very little we can regard the observations contained in each group as being sufficiently accurate. To proceed thus, we must have a considerable number of observations, and they must not contain very discordant measures. If the partial means do not thus agree, the precision of the general mean is very doubtful.

Example. In the observation on the temperature of Providence, R. I., by Prof. A. Caswell, during twenty years, the mean temperature of the first ten years is . . . . . . 48°·1 F.
" " second " " " . . . . . . 48°·3

Difference, ... 0°·2
The mean of the whole twenty years is . . . . 48°·2

This shows that one decade is sufficient to give the mean temperature to within about one-tenth of a degree.

It often happens that we find the measures grouped together in series, depending on the periods or on the various circumstances when they were made. It is then indispensable to calculate the partial means of these series, and if there has existed, during the time of the observations, a constant cause of errors, but of variable intensity, it may manifest itself by taking, as above, the means of several groups.

Errors of Observation, called also *residual errors*, are the differences between each particular measure, and the mean of all the measures. These errors are always very small when we operate with care.

The *absolute value* of the quantity sought remaining always unknown, we substitute for it the mean, in the calculation of errors, without which it would be impossible to appreciate them.

When the errors are purely accidental it is found that they can be arranged according to a most remarkable law. If they are grouped by their signs and according to their magnitude, we find that the positive errors and the negative errors are equal in number and in aggregate value, and that they diminish rapidly as we go from the mean, according to a regular law; in other words, the smallest errors are the most frequent and are principally accumu-

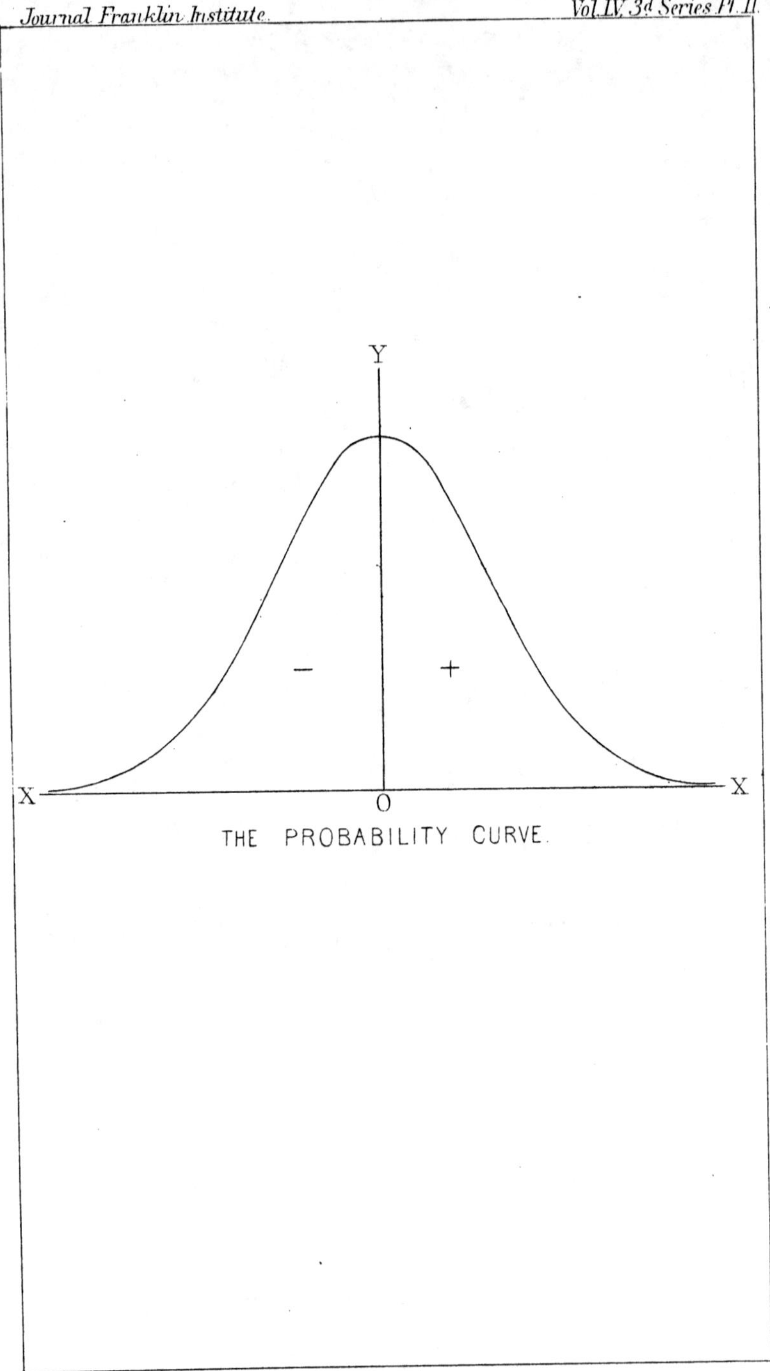

THE PROBABILITY CURVE.

lated around their mean value. This remarkable fact is shown in the figure, where, on the left of the axis O Y are placed the differences or errors of observation less than the mean, or —; and on the right those greater than the mean, or +.

Theoretically, this curve, called the *curve of probability*, should be perfectly symmetric and regular. If it departs much from this form it shows that the observations or experiments have not been made with sufficient care, or that their number is insufficient, or that there exists a permanent cause of error arising either from the observer himself (personal equation), or from the manner of operating.

Suppose we have now found a series of measures which, being divided into groups, give as many means which differ by little; that the difference of each mean from the general mean is very small; and, finally, that these errors are distributed about the mean according to the curve of probability; then the cause of the errors is purely accidental, they tend to diappear from the mean, and the *theory of probabilities* gives us the numerical value of the *degree of precision.*

The first principle of this theory is, that *the precision in the mean increases as the square root of the number of observations from which it is derived.* Thus, the circumstances in both series of observations remaining the same, from sixteen observations we have in favor of the precision of the mean, a probability twice as great as for four observations.

We must now consider another element, which is, the error of each partial measure, or its departure from the mean.

We call the *mean error* not the arithmetical mean of the errors, but *the square root of the mean of the squares of the errors.* Let ε, ε', ε'', &c., ... be the errors of observation, their mean error, e, will be—

$$e = \sqrt{\frac{\varepsilon^2 + \varepsilon'^2 + \varepsilon''^2 + \quad \cdot \cdot \cdot \cdot \cdot \cdot}{n}}$$

n being the number of observations.

By means of this *mean error* we can calculate what is called *the probable error*, either of the mean or of a single observation.

The probable error is that quantity which has such a magnitude that there is the same probability of the error, in the quantity determined, being greater as there is of its being less by this quantity.

It is demonstrated that the probable error of a single observation is—

$$\frac{2}{3} e.$$

The probable error of the mean result is—

$$E = \frac{2}{3} \frac{e}{\sqrt{n}}.$$

We thus see that it diminishes in the inverse ratio of the square root of the number of observations.

We will now show from examples, taken from the appendix to Gerhardt's *Traité Elémentaire de Chimie*, how the foregoing principles and formula can serve to indicate the exactness of a series of measures.

M. Dumas made experiments to determine the composition of water, and to verify the theoretical law of Prout, that all the chemical equivalents are exact multiples of that of hydrogen. The equivalent of oxygen being represented by 100, that of hydrogen should be exactly 12·5. But a series of nineteen experiments gave, after the corrections were made, the following numbers, which we arrange according to their magnitude.

12·472	12·508
·480	·522
·480	·533
·489	·546
·490	·547
·490	·550
·490	·550
·491	·551
·496	·551
	·562

The mean of the numbers is 12·515; the sum of the squares of the errors is 0·0173; the mean error $e = 0·030$; the probable error of a single observation is 0·020, and the probable error of the mean result $E = 0·0046$. It would appear from this that an error of 0·015 is impossible, and that the equivalent of hydrogen ought to be above 12·500.

But if we observe attentively the table, we will see that the numbers, instead of being accumulated, according to the law of probability, around the mean, tend rather towards the extremes. The means of the two columns are 12·486 and 12·542, which differ too much to accept the hundredths when we unite them. It therefore follows that the method of experiment by which were obtained the numbers, is too gross to decide that the equivalent of hydrogen departs from 12·5, and we cannot therefrom decide against the law of Prout.

We see from this example, taken from *The Theory of Chances*, by M. Cournot, how important it is, before drawing our conclusions, to examine with care the individual observations whence we deduce the mean, so that we may satisfy ourselves of their exactitude.

We often exaggerate the accuracy attained, which is necessarily limited. A quantity cannot be estimated to an indefinite precision even when the observations are indefinitely extended. In the measure of a length, for example, whatever care we may take we can seldom surely rely on more than five figures. Thus, one of the base-lines in the triangulation of France was found to be 6075·9 toises; we find that there can easily occur in this number an error of 0^{t}·1. In the estimation of weights there is also a limit of precision which we cannot surpass, whatever may be the accuracy of the balance and the skill of the operator. Here also we cannot rely on more than five significant figures.

Still greater reason is there for care in assuming a certain precision of result when the measure is the result of several different physical operations, each of which brings with it its own error. In the determination of the number $g = 9^{m}$·809, which represents the intensity of gravity at Paris, (or the velocity acquired by a body after falling one second,) the four figures which we have given are those only on which we can rely, and it would be altogether illusory to extend the decimal. If we compare the numbers found by able observers, we find that they differ even in this fourth figure.

Intensity of gravity at Paris.	Length of second's pendulum.
Borda, 9^{m}·8089	0^{m}·99385
Biot, 9 ·8091	·99387
Bessel, 9 ·8094	·99390

For the length of the second's pendulum we should take 0^{m}·9939, but remembering that we can barely rely upon the result to $\frac{1}{10}$th of a millimetre.

Physicists and chemists have often to estimate the exactness it is possible to attain in the determination of densities and equivalents. By repeating these operations several times in succession, by varying the methods, by taking different specimens of the same body, and finally, by submitting the results to the numerical tests which we have indicated, we arrive at the degree of precision of the results.

We thus find that the number of *certain* figures in this class of determinations is less than we would *à priori* have supposed, and that no more reliance is to be placed in several of the decimals that are usually retained, than if we wrote them down at random. We can often discover this influence of chance in a series of measures. and distinguish those decimal figures which we ought to suppress as being altogether arbitrary, since they are less than the errors of observation. Suppose that we estimate a length which can only be appreciated to a millimetre; if we retain the tenths of millimetre, they will be entirely of an arbitrary value. The figures written in the tenths will be irregular in the successive measures, absolutely, as if we obtained them by drawing at random, out of our pocket, marbles having pasted on them the figures 0, 1, 2, 3, 4, 5, 6, 7, 8, 9. But, as the mean of these ten figures is 4·5, it follows that if we have a great many results, the mean of the figures of tenths of millimetre will be exactly 4·5. And *vice versa*, if in a long series of numbers, given by observation, we find that the mean of all the decimal figures of the last order is 4·5, we can omit them. If it is the same with the figures of the next higher order, we suppress them also and so on.

M. Schrötter determined the equivalent of phosphorus, by weighing the phosphoric acid produced in the combustion of a known weight of that element. These experiments were made upon amorphous phosphorus, previously dried at 150° C., in an atmosphere of carbonic acid or of hydrogen. The transformation of the phosphorus into phosphoric acid, was effected in a combustion tube, through which passed a current of perfectly dry oxygen. After the combustion, the phosphoric acid was sublimed in an atmosphere of oxygen, so as to oxydise any traces of phosphorus acid which might have been produced.

The following are the results:—

Phosphoric acid produced by one part of phosphorus.	2·28909
	·28783
	·29300
	·28831
	·29040
	·28788
	·28848
	·28856
	·28959
	·28872
Mean, ...	2·28919

We find that the sum of the figures of the last column on the right is 46, the mean of which is 4·6, since there are ten results. In the same manner the mean of the next column, to the left, is 4·4. We therefore retain only the first three decimal figures, the remaining two being solely due to hazard. This done, if we calculate the error of each observation, or the departure from the mean, we will find that the sum of the squares of these errors is 0·000022; the mean error $e = 0·0015$, the probable error of a single observation is 0·001, and the probable error of the mean is $E = 0·0003$.

We can therefore admit that one part of phosphorus gives 2·289 of phosphoric acid, and that the error of this result certainly does not exceed six times the probable error, that is to say, 0·002. The corresponding equivalent of phosphorus is 31·027, and its probable error = 0·004, which is expressed thus—

$$31·027 \pm 0·004.$$

But that supposes that the analyses are only effected with accidental errors, while it may readily happen that they are subject to a constant error. It would then follow that the figure of the hundredths could not be regarded as certain; therefore M. Schrötter adopts simply 31 as the final result of his researches.

It is from the constant errors existing that we may err in the conclusions that we prematurely draw from the calculation of the mean of the probable errors. And it often happens that subsequent observations show that the mean adopted at first is affected with an error far exceeding the probable error which we attributed to it.

It is a characteristic of the arithmetical mean, that it makes *the sum of the squares of the residual errors a minimum.*

To illustrate this principle geometrically, suppose that several observations, made to determine the position in space of a point, A, give several positions around the true point, A. Connect the points given by observation by straight lines; then the centre of figure or centre of gravity of the polygon, which is formed by joining these points, is evidently the most probable position of A. Draw lines from the angles of the polygon to its centre of figure; then these lines will represent the several errors of the observations. Now, it is a property of the centre of gravity of such a polygon, that the sum of the squares of these lines is less than the sum of the squares of similar lines drawn to any other point within the polygon. Therefore, *if the sum of the squares of the errors is the least possible, we have the nearest mean we can obtain from the observations.*

This is the fundamental principle of the celebrated theorem of *Least Squares*, of Gauss and of Legendre, which has rendered such inestimable service to physics and astronomy.

Suppose that, instead of having to determine, as in the above examples, a single unknown quantity—which is the mean of the several observations,—we propose to determine several unknown quantities entering as functions in equations. As observations or experiments give the known quantities of these equations, we will have as many equations as there are observations. Having, for example, two quantities to be determined, x and y, suppose that we have made three observations, giving three equations. The values of x and y deduced from two of these equations, will not generally satisfy the third. The difficulty consists in using all the equations at once, so that we can deduce from them a series of values which will satisfy, in the most accurate manner possible, all the equations. The theory of the method of least squares shows, in order that this may happen, that the sum of the squares of the errors must be a minimum.

In a series of observations or experiments, let us suppose that the errors committed are denoted by e, e', e'', &c., and suppose that by means of the observations, we have deduced the equations of condition—

$$\left.\begin{array}{l} e \;\; = h + a\,x + b\,y + c\,z + \&c. \\ e' \;\, = h' + a'\,x + b'\,y + c'\,z + \&c. \\ e'' = h'' + a''\,x + b''\,y + c''\,z + \&c. \\ e''' = h''' + a'''\,x + b'''\,y + c'''\,z + \&c. \\ \qquad\quad \&c. \qquad \&c. \qquad \&c. \end{array}\right\} \quad (1.)$$

Let it be required to find such values of x, y, z, &c., that the errors e, e', e'', e''', &c., with reference to all the observations shall be the least possible.

If we square both members of each equation, in group (1), and add them together, member to member, we shall have—

$$e^2 + e'^2 + e''^2 + \text{\&c.} = x^2 (a^2 + a'^2 + \text{\&c.})$$
$$+ 2x \left\{ (a h + a' h' + \text{\&c.}) + a (b y + c z + \text{\&c.}) \right.$$
$$\left. + a' (b' y + c' z + \text{\&c.}) + \text{\&c.} \right\} + \text{\&c.},$$

an equation which may be written—

$$e^2 + e'^2 + e''^2 + \text{\&c.} = u = P x^2 + 2 Q x + R + \text{\&c.}$$

Now, in order that $e^2 + e'^2 + \text{\&c.}$, or u, may be a minimum, it is necessary that its partial differential co-efficients, taken with respect to each variable, in succession, should be separately equal to O. Hence,

$$\frac{d u}{d x} = p x + q = 0, \text{ or } x (a^2 + a'^2 + \text{\&c.})$$
$$+ a h + a h' + \text{\&c.} + a (b y + c z + \text{\&c.})$$
$$+ a' (b' y + c' z + \text{\&c.}) + \text{\&c.} = O;$$

and similar equations for each of the other variables. Hence, we deduce the principle, that, in order to form an equation of condition for the minimum, with respect to one of the unknown quantities, as x for example, we have simply to multiply the second members of each of the equations of condition by the co-efficient of the unknown quantity in that equation, then take the sum of the products, and place the result equal to O. Proceed in this manner for each of the unknown quantities, and there will result as many equations as there are unknown quantities, from which the required values of the unknown quantities may be found by the ordinary rules for solving equations.

Let it be required, for example, to find from the equations—

$$\left.\begin{aligned} 3 - x + y - 2z &= O \\ 5 - 3x - 2y + 5z &= O \\ 21 - 4x - y - 4z &= O \\ 14 + x - 3y - 3z &= O \end{aligned}\right\} \quad (2),$$

such values of x, y, and z, as will most nearly satisfy all of the equations.

[It is to be remarked, that we must necessarily admit that these equations already approximate closely in value, that is to say, that the values of x, y, and z deduced from the three first will satisfy approximately the third; without this, the problem is absurd, and the observations which have given these equations are unworthy of confidence.]

"Following the rule, and multiplying the first member of each equation by the co-efficient of x in that equation, we get the products—

$$\begin{array}{l} -3+x-y+2z \\ -15+9x+6y-15z \\ -84+16x+4y+16z \\ 14+x-3y-3z, \end{array}$$

and placing their algebraic sums equal to O, we have—

$$27x+6y-88=O, \qquad (3.)$$

Proceeding in like manner with respect to the unknown quantities y and z, we obtain the equations—

$$6x+15y+z-70=O, \qquad (4.)$$
$$y+54z-107=O, \qquad (5.)$$

Combining equations (3), (4), and (5), we find—

$$x=2{\cdot}4702,\ y=3{\cdot}5507, \text{ and } z=1{\cdot}9157,$$

which most nearly satisfy all of the equations in group (2).

To show the practical application of this principle, we will suppose that it is required to investigate the values of a constant in an equation, by means of several independent experiments.

It is demonstrated from theory that the length of the pendulum which beats seconds, in any latitude, is given by the formula—

$$L=x+y\sin.^2 l, \qquad (6,)$$

in which L denotes the length of the pendulum; l the latitude of the place on the surface of the earth, and x and y are constants to be determined. In consequence of errors incident to observation, the values of x and y cannot be accurately determined by means of a single observation; taking the *metre* (denoted by m) as the unit of measure, suppose that the length of the second's pendulum has been measured in six different places, whose latitudes are known, and that the following equations have been deduced:

$$\left.\begin{aligned}
e' &= x + y \times 0^{m}.3903417 - 0^{m}.9929750 \\
e'' &= x + y \times 0^{m}.4972122 - 0^{m}.9934620 \\
e''' &= x + y \times 0^{m}.5667721 - 0^{m}.9938784 \\
e^{iv.} &= x + y \times 0^{m}.4932370 - 0^{m}.9934740 \\
e^{v.} &= x + y \times 0^{m}.5136117 - 0^{m}.9935967 \\
e^{vi.} &= x + y \times 0^{m}.6045628 - 0^{m}.9940932
\end{aligned}\right\} \quad (7.)$$

Applying the rule already deduced, to these questions, we find the equations—

$$6x + y \times 3^{m}.0657375 - 5^{m}.9614793 = 0, \quad . \quad . \quad (8),$$

$$x \times {}^{m}.0657375 + y \times 1^{m}.5933894 - 3^{m}.0461977 = 0, \quad (9.)$$

Combining equations (8) and (9), we obtain—

$$x = 0^{m}.9908755, \; y = 0^{m}.0052942.$$

Substituting these in equation (6), it becomes—

$$L = 0^{m}.9908755 + 0^{m}.0052942 \, sin.^{2} \, l, \quad . \quad . \quad (10.)$$

By means of formula (10), the length of the second's pendulum may be found, by computation, at any place whose latitude is known. In like manner, the method of least squares may be applied to a multitude of similar cases." (See art. *Least Squares*, *Dict. of Mathematics*, by Davies and Peck, N. Y., 1865.)

It is impossible, in lectures of this character, to give more than an idea of this fertile principle of *Least Squares.* The reader is referred to the following works for a full discussion of the method. *Méthode des moindres carrées; Mémoires sur la combinaison des observations, par Ch. Fr. Gauss, traduit par J. Bertrand*, Paris, 1855; also *Method of Least Squares*, by Prof. William Chauvenet, in the appendix to his *Spherical and Practical Astronomy*, Phila., 1863. From this work the figure of the probability curve is taken.

2. *Method of Multiplication.*

The method of multiplication consists in measuring a known multiple, by n, of the quantity to be determined, and then dividing the result by n. By this method the error, committed in the direct measure, is divided by n.

Example 1. To find the diameter of a fine metallic wire, we wrap a certain number of turns, say 200, of the wire on a metal cylinder, taking care to press the coils until all are in contact. Measure the length of the 200 coils, and dividing this length by 200, we have—supposing all the coils are formed of wire of the same diameter—the diameter of the wire to the $\frac{1}{200}$th of the error committed in the measure of the length of the 200 coils. Suppose, for example, we have made an error of $\frac{1}{20}$ inch in measuring the length of 200 coils; then the error in the determination of the diameter of the wire is $\frac{1}{200}$ of $\frac{1}{20}$ or the $\frac{1}{4000}$th of an inch.

Example 2. By a similar operation we determine, with great accuracy, the distance between the contiguous threads (called *the pitch*) of a micrometer screw.

Example 3. Saxton's Reflecting Comparator is a beautiful illustration of the principle of multiplication, in which the direct measure is a very large multiple of the quantity whose value we desire.

Example 4. We may obtain a very accurate determination of the weight of a body, with a balance with equal arms, by equilibrating the body with shot, &c., placed in the other pan; then placing the body in the same pan with the shot, &c., we obtain in the other pan, in shot, *twice* the weight of the body. We now add the shot of twice the weight of the body, to the pan containing the body, and on again equilibrating, we have in the pan opposite the body, a weight of shot equal to *four times* the weight of the body, and so on. We finally have in a pan a known multiple, by n, of the weight of the body, which, divided by n, gives the weight we seek.

Example 5. *The Method of Repetition.*—Describe the method with the aid of diagrams. It is a very ingenious application of the method of multiplication to the measurement of angles, by means of the *repeating circle*, invented by Tobias Mayer, of Göttingen, in 1762, and subsequently improved by Borda, of Paris.

3. *Method of Successive Corrections.*

This method, used frequently in astronomy, is also often employed in researches in Physics. We will explain it by an example (from *Daguin's Traité de Physique*, vol. i., page 30). Let it be proposed to determine the capacity of a ground glass-stoppered flask. If we knew the weight of the water at 4° C., which the flask holds, then just as many grammes as there are in this weight, are there cubic centimetres in the flask; for a gramme is the weight of one cubic centimetre of pure water at 4° C. It is, therefore, required to determine the weight of the flask-full of water at 4° C., which we suppose is the temperature when the experiment is made. We first weigh the flask, full of air, then full of water at 4° C. Let p and P be the two weights, respectively.

P—p will represent the weight of the water, if the flask, during the first weighing, had contained no air, whose weight is added to that of the flask. The value P—p is therefore too little by the weight of that quantity of air. Suppose that the water at 4° C., weighs n times as much as an equal volume of air, taken at the same atmospheric pressure and temperature as in the experiment. We will see further on how this number n is obtained. The weight of the air which fills the flask will then be $(P-p)\frac{1}{n}$, if we suppose for the moment that P—p is the exact weight of the water. Then the weight of the water which fills the flask will be—

$$(P-p)+(P-p)\frac{1}{n}=(P-p)\left(1+\frac{1}{n}\right), \quad . \quad . \quad (1.)$$

This expression, however, does not represent the exact weight of the water, because the term $(P-p)$, and consequently the term $(P-p)\frac{1}{n}$, are too small. But the error which exists in value (1) is less than that which affects the value P—p. If we now employ the value (1) to calculate the weight of the air, this weight will be—

$$(P-p)\left(1+\frac{1}{n}\right)\frac{1}{n};$$

and the weight of the water will become—

$$(P-p)+(P-p)\left(1+\frac{1}{n}\right)\frac{1}{n}=(P-p)\left(1+\frac{1}{n}+\frac{1}{n^2}\right), \quad . \quad (2;)$$

a value which is a still greater approximation to the true value, and which, multiplied by $1:n$, will give the weight of the air with still greater precision.

By adding this weight to P—p, we will have for the weight of the water—

$$(P-p)\left(1+\frac{1}{n}+\frac{1}{n^2}+\frac{1}{n^3}\right);$$

nearer than any of the preceding values. By continuing in this manner, we can carry the approximation as far as we desire.

§ IV. *Manners of Expressing a Law—Law evolved from the numerical results of Observations and Experiments.*

The first step in that investigation, which has for its object the discovery of the law of a class of facts, is the minute examination and description of the phenomena and of the circumstances which accompany them, and the determination of those conditions necessary for their production. This having been accomplished, we observe that there always exists between the different circumstances of an associated class of facts, relations or dependencies which bind them together in such a manner, that if we change one of the circumstances of the phenomenon, the others experience determinate modifications.

For example. In the compression of air, we have seen that the smaller the volume into which the air is forced, the greater the force of compression required; and on further examination with *measurements* taken of (1) the volumes occupied (2) under different pressures, we find that the relation which binds (1) and (2) is that *the volumes of the air are inversely as the pressures to which it is subjected.*

Such an expression of the relation existing between the different circumstances of a class of associated facts is, as we have seen, (§ I.) *a physical law.*

"These laws rule all phenomena, and are their most complete representation. Their existence did not escape the acute minds of the philosophers of antiquity. Plato, questioned concerning the occupations of the Deity, replied that *He geometrized without ceasing;* wishing thereby to express according to Montucla, that the universe is governed by geometrical laws" (DAGUIN'S TRAITE DE PHYSIQUE, Vol. I., p. 8).

If the law is the expression of a general *quantitative relation* existing between the associated facts of the phenomenon, it can be replaced by a line referred to coördinate axes, as in the method of Analytical Geometry.

Example 1. Thus, *a Parabola* represents the law of falling bodies, because the origin of rectangular coördinates being at the principal vertex, the abscissas are to each other as the squares of their corresponding ordinates; or,

$$x' : x'' :: y'^2 : y''^2;$$

and the law of falling bodies is that *the spaces fallen through are as the squares of the times;* therefore, if any ordinate represent in units of length the units (seconds) of time occupied in the fall, the units of length contained in its corresponding abscissa will give the units of length fallen through; one unit of length in this case—being a function of the intensity of gravity—is, for the latitude of New York, equal to 16 feet 1 inch.

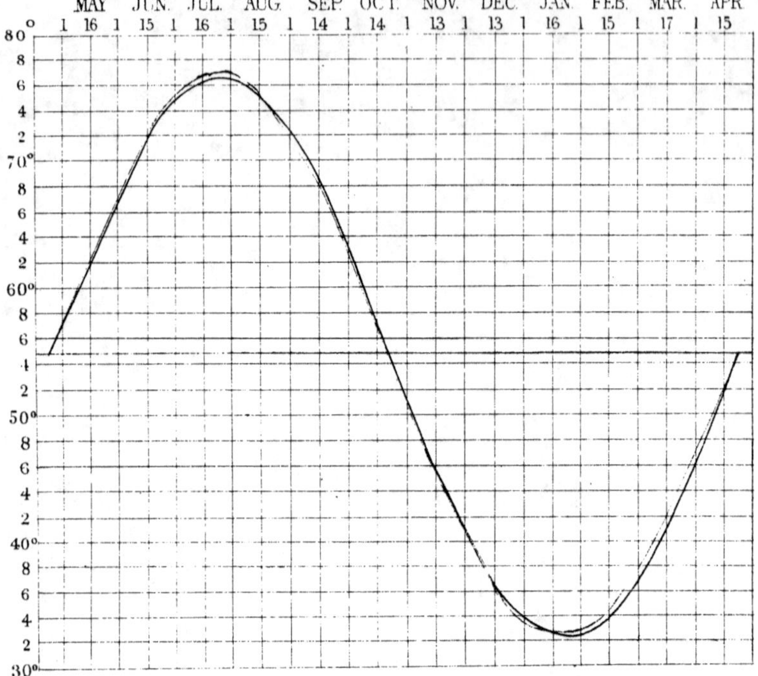

THERMAL CURVE OF BALTIMORE WITH CORRESPONDING SINUSOID.

Fig. 1

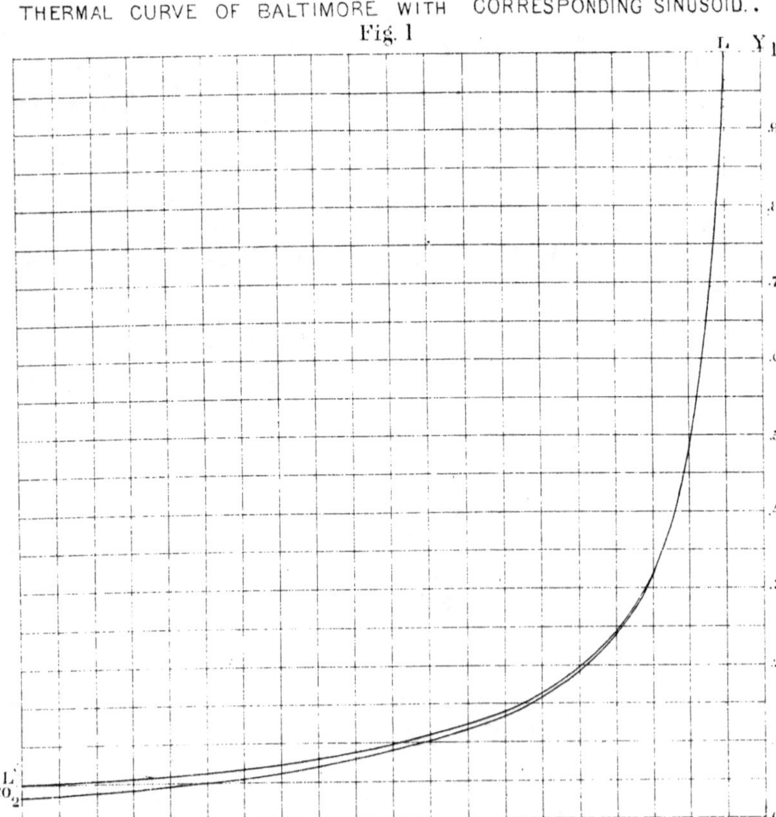

The above is directly shown in MORIN'S MACHINE, where the body describes the parabola by falling parallel to the axis of a uniformly revolving cylinder, against which a pencil, attached to the falling body, gently presses. The curve thus formed is found on spreading out the paper, which covers the cylinder, and measuring its abscissas and ordinates, to be a parabola. Thus, in this beautiful experiment, the body itself writes on the paper the law of its motion.

Example 2. The law of the compression of gases is given by a curved line (see Fig. 1, Plate III. Curve L L′), whose abscissas representing pressures and ordinates volumes, has for its equation

$$y=\frac{1}{x}.$$

This curve is an equilateral hyperbola, and therefore the axes of X and of Y are asymptotes to the curve; and this should be so, for the gas, supposed to be non-liquifiable, has under an infinitely great pressure, still a definite volume, which is expressed by the ordinates corresponding to that pressure. If, however, the gas is not permanent (like carbonic acid), the curve will cut the axis of X at a point corresponding to the pressure producing liquifaction. If all those gases which are susceptible of liquifaction—though the pressures required for this result for certain gases might be beyond the limit of practicable experiments—had curves whose deflections towards the axis of X all followed the same law, we could, by projecting the curve of any gas, as far as the limit of experiment, determine at what point it would cut the axis of X, and thus determine the pressure of liquifaction.

See Fig. 1, Plate III. The curve L L′ represents Mariotte's law, that the volumes of gases are the reciprocals of the pressures; while L CO_2 shows the fraction of a unit of volume, measured on the axis O Y, occupied by carbonic acid gas under pressures, from 1 up to 20 atmospheres, measured on the axis O X. If the curve of hydrogen were projected on this diagram, it would sweep above L L′, departing very slightly from this line on the side opposite the curve, L CO_2.

The advantage of the graphic method of expressing a law is, that it represents *to the eye* the *continuous* relation between the associated facts of the phenomenon, and therefore gives, by direct measurement, the quantity answering to any measure taken of either of the related magnitudes embraced in the projected curve. Also, by the mathematical discussion of the curve, new relations which otherwise might pass unknown, may be evolved.

Sometimes the relation existing between the associated facts is too complicated to be susceptible of a concise quantitative statement. In that case, we lay off on an axis of rectangular coördinates lengths of abscissas equal to various values of one of the quantities, and on their corresponding ordinates lengths equal to the corresponding values of the other quantity. We then draw *libera manû*, through the several points thus determined, a curved line, and this line will be the *continuous expression* of the relation existing between the two quantities considered.

It may happen that the inspection of the curve leads to the discovery of a law, which we could not have made from the direct comparison of the numbers given by experiment or observation. For example, if the curve was one of those lines studied by geometers, and whose properties are well known, the relation which exists between the abscissas and ordinates of the curve, or, in other words, its equation (referred to the axes X and Y), will express the law which we seek. If, therefore, we recognize by mere inspection, that the curve resembles a known line, we must proceed in the following manner to the verification of that supposition. We write the general equation of the curve, referred to the axes X and Y, giving it indeterminate co-efficients; we then successively substitute in that equation as many values of x and of their corresponding ordinates y, as there are co-efficients; which gives *equations of condition*, by means of which we calculate the values of these co-efficients, which we then substitute in the general equation. We then successively substitute for x and y other numerical values given by the experiments or observations, and we examine if these numbers satisfy the equation. If this is the case, we conclude that the curve is really the one we suspected, and its equation expresses the mathematical law of the phenomena.

For a simple example, take the curve described by Morin's Machine for showing the law of falling bodies. If, from preliminary measurements, we suspect the curve to be a parabola, we have, for the general equation of that curve—

$$y^2 = 2\,p\,x;$$

in which, placing x' and y' from measured coördinates of the curve, we have—

$$2\,p = \frac{y'^2}{x'}$$

substituting this value of the parameter in the general equation, and making x and y successively x'' y'', and x''', y''', &c., obtained from other measurements on the curve, we find that—

$$y''^2 = 2\,p\,x''\,;\ y'''^2 = 2\,p\,x''', \text{ \&c.}$$

whence we conclude that the curve is the quadratic parabola, and therefore—

$$x' : x'' :: y'^2 : y''^2\,;$$

and since in the curve described by this machine, x' and x'' are equal to the spaces fallen through by the body in the corresponding times y' and y'', we have the law that *the spaces fallen through are as the squares of the times.*

Example, in which more complex relations exist—

We have (in a paper preparing for publication), projected the thermal curves, of the variation of temperature from day to day throughout the year, of fifteen places, differing in latitude, longitude, elevation above and distance from the sea, and in other topographical features; and on discussing these curves, we find that they can all be replaced by *sinusoids*, differing from each other only in the amplitude of their flexures, which corresponds to half the annual range of temperature of the different stations.

Lengths of abscissas standing for days and lengths of ordinates representing degrees of temperature (see Plate III., Thermal Curve of Baltimore, with corresponding sinusoid; where the horizontal red line marks the mean annual temperature, while the curved red line is the annual thermal variation, and the curved black line its corresponding sinusoid), we obtain a series of points, by laying off on ordinates erected from abscissas corresponding to certain days, the mean temperatures of those days, and then drawing a curved line through these points, we have the annual thermal curve. This curve has its origin at the point of the axis of X which corresponds, in the average, to the date of the 21st of April; the mean temperature of that day being the mean temperature of the year.

(Diagrams of thermal curves, with corresponding sinusoids of fifteen stations exhibited.)

Now projecting on the line of mean annual temperature of each curve, as the base, *a sinusoid* the length of whose base (=360°) equals the length of the year; whose point of origin is on the base at the point corresponding to the 21st of April, and whose amplitude equals the half yearly range of temperature, we have, in every instance, so close a coincidence between the thermal curve and the sinusoid, that we can confidently say, that the variation of temperature throughout the year follows the variation of the sinusoid curve.

The equation of the sinusoid is—

$$y = sin.\ x;$$

and as the sines of the first and second quadrants are + and of the third and fourth —, it follows that this is a recurring curve, and up to *sin.* ($x = 180°$) the flexure is above the base line, while from *sin.* ($x = 180°$) up to *sin.* ($x = 360°$) the curve lies below the line, and so on; each succeeding 180° being on the opposite side of the line of mean temperature from the preceding.

Therefore, expressing any date in days reckoned from April 21st, and converting these days into lengths of arc $= x$ at the rate of one day $= \frac{360°}{365} = 0°\ 59'\ 1''$, and making the sin. 90° or R $= \frac{1}{2}$ the mean yearly range of temperature, we can readily determine the mean temperature of any date of the year.

1. *Problem.* What is the mean temperature of the 10th of June, at Paris?

The mean annual temp. from 46 years observations,	= 51·° 3 Fah.
Half the yearly range, ...	= 15·° 47
Days from 21st April to 10th June, . . .	= 50 days.
The above number of days in arc, . . .	= 49° 10′ 50″
Natural sine of above arc, ...	= ·756773
Radius in this example, ...	= 15·47
Then the temperature of 15th June equals ·756773× 15°·47 + 51°·3	= 62°·90
Mean temperature of 15th June, from observations of 46 years, ...	= 62°·99
Difference, . .	= 0°·09

2. *Problem.* What is the mean temperature of the 13th of Dec. at Baltimore? The temperature from *observations* = 35°·6. Data for solution. From 35 years observations, the mean annual temperature = 54°·4. The yearly range = 44°·4.

3. *Problem.* Calculate the mean temperature at Baltimore, of of the 14th October. The observed mean temperature of this day is equal to 56°·5.

4. *Problem.* Determine which is the warmest and which is the coldest day in the year at St. Petersburg. Data for the solution. Mean temperature, 38°·75. Annual range, 50°.

From observations of 30 years, the warmest day at St. Petersburg is on the average, the 25th July; the coldest, the 19th January.

We can therefore determine the mean temperature of any day of the year for any station (not situate in the tropics), whose mean temperature and yearly range of temperature are known. The mean temperature of any place can be found by taking the temperature once a month during one year, of any spring which issues from a depth of a few feet below the surface of the ground. Thus, the temperature of a spring in Baltimore, from the mean of one observation a month, during one year, was 54°·25, only 0°·15 below the mean given by 35 years observations of the thermometer in the air.

If a law could be established connecting different parallels of latitude with the yearly range of temperature, we could arrive at the other datum, and then the curve of annual variation could be projected from a knowledge of the latitude and longitude, with the determination of the mean temperature of a spring.

This method of representing to the eye, by means of curved lines, the nature of the laws concealed within a mass of numbers, presents the advantage of representing the continuous and periodical nature of cosmical action. It also presents to the eye *a form*, easy to seize and remember, instead of an abstract statement referring to relations of quantities, difficult to be conceived and sometimes impossible otherwise to discover. Thus the thermal curve of Berlin is the graphic expression of the relation of the 365 days of the year to observations made three times a day for 110 years, or, to 120,450 separate numerical quantities.

The curves show by their deformities, either that our data are not the results of sufficiently extended observations to give, by their projection, the expression of a law, or, that another cause is acting which increases or diminishes the effect due to the cause whose rule of action is expressed, in the main, by the curve.

Thus, in the thermal curves of Greenwich, of Paris and of Rome, an upward deflection exists, from the middle of January to the middle of April, causing a departure of about 2° Fahr. from the sinusoidal curves. This departure, at first sight, would appear as opposed to the assumption that the annual variation of temperature follows the variation of the sinusoid; but, in reality, it is "an exception which proves the law;" for it is found that during those months the prevailing winds for those stations are W. S. W., and Dove has conclusively shown that in the northern hemisphere, the thermometer stands highest, on an average, with a S. W. wind. This deflection, however, does not occur in the curve of Berlin, and there a S. 17° W. wind prevals during those months. But this is easily accounted for, and the explanation will at the same time give a sufficient reason for the upward deflection of 2° in the thermal curves of Greenwich, Paris and Rome.

The prevailing wind of Greenwich from the middle of January to the middle of April, reaches that station after having traversed the Atlantic ocean, whose surface water has, during the above months, even in the latitude of the English Channel, an average temperature of 52° Fahr. Now, one cubic foot of water in cooling 1° Fahr. will give out heat sufficient to raise 3080 cubic feet of air 1° Fahr., and the wind after traversing this surface of a liquid, of high specific heat, 13°·5 above the mean temperature of the air during February, passes over only about 85 miles of lowlands before reaching Greenwich. The same prevailing wind reaches Paris after having also passed over the Atlantic,—of even a slightly higher temperature than in the case of the Greenwich wind,—and traversed about 250 miles of land-surface not likely to affect its temperature, since this very land has received its thermal condition mainly from the supernatant air; but the S. 17° W. wind reaches Berlin after having crossed the snow covered summits of the Appenines and Alps, and 750 miles of land.

This general graphic method of expressing the relation which pervades a class of facts, evidently gives a ready means for *interpolation*, and even where *deformities* exist, a mean or true curve can be obtained by *sweeping* a line through the mean position of the irregularly placed points given by the observations or experiments. By this last method, Sir John Herschel determined the orbits of several of the double stars (*Transactions of the Royal Astronomical Society*, Vol. V., 1832), and M. Regnault (*Mémoirs de l'Institut*, *t.* 21, 1847, page 316, *et seq.*, and page 574, *et seq.*), from his measures of the tension of vapor of water corresponding to different temperatures, succeeded in ascertaining the extreme oscillations of the errors of those measures, by drawing a mean curve through the sphere of the points given by his experiments. He then substituted this curve for the numbers directly determined by experiment, because this curve expressed the relations of those numbers practically corrected of their errors.

It should be remembered that each number at which we finally arrive, before we place it down as a point to serve in the projection of the curve, is *the most probable mean* which we can obtain from the discussion of the numbers from which it is derived, (see § III.), and that each of these points so determined is arrived at independendently of any other point, and therefore the graphic method is especially adapted to combine all of these independent determination of points in one curve, and to make them mutually correct each others errors.

The curve thus obtained can be rendered more serviceable in practice, by obtaining from it, if possible, an equation which will express it; but if we find that it cannot be expressed by a single equation, we can, by the aid of *formulæ* of *interpolation*, find the value of each ordinate corresponding to as small an increase in each successive abscissa as we desire. We should not, however, apply a formula of interpolation to the numbers before we have obtained from them a mean curve. The interpolation is merely used to give a ready means of obtaining the values of any related magnitudes expressed by a curve, which shows the relations of the numbers corrected of their errors; for, as M. Regnault remarks, "the graphic method when it is properly executed, is preferable to all methods of interpolation by calculation; it permits us to distinguish, immediately, the variations due to the accidental errors of the observations, and the constant errors which depend on the diversity of the methods which we have employed."

§ V. *The General Properties of Matter.—The Constitution of Matter according to the Molecular Hypothesis.*

MATTER is that which affects our senses. (See § I).

A body is a portion of matter limited in all directions.

In the different manners in which bodies affect our senses, consist the properties of bodies. These properties are either *general*, that is, belonging to all bodies, whether solid, liquid or gaseous, and therefore come under the head of physics, or they are *special*,

7

and therefore belong to the province of chemistry. (See definitions of Physics and Chemistry, in § I).

Matter possesses several general properties, of which two are called *essential* properties; for without them we cannot conceive of the existence of matter, and therefore they can serve to define it. These two properties are *extension* and impenetrability.

The following is a list of the general properties of bodies:—

1. Extension. } Necessary to our perception of matter.
2. Impenetrability. }
3. Figure.
4. Divisibility.
5. Compressibility.
6. Dilatability.
7. Porosity.
8. Mobility. } Ultimate properties according to the molecular hypothesis.
9. Inertia. }
10. Attraction and }
11. Repulsion. }
12. Polarity.
13. Elasticity.

1. *Extension.*

Extension is the property which every body possesses of occupying a portion of space, which we call its volume. A body always presents the three dimensions—length, breadth, and thickness; and it is by abstraction only that in geometry we consider *surfaces*, which are the boundaries of bodies, and have only length and breadth; and *lines*, which are the boundaries of surfaces, and have only length; and *points*, which are the terminations of lines, and have alone position.

For the measurements of extension, see § II.

2. *Impenetrability.*

The property which every body has of excluding every other body from the space which it occupies.

In ordinary language, we say that one body is penetrated by another, but in all these cases it is found that the particles of the one body are merely *displaced* by the other.

Examples. A needle *penetrating* (displacing) mercury contained in a fine glass tube. The mercury rises in the tube, as the needle descends, to an amount exactly equal to what would be produced by pouring into the tube a quantity of mercury equal to the volume of that portion of the needle below the surface of the mercury.

A liquid cannot be poured into a vessel unless the air, which it contains, goes out as it enters.

Consider the images formed in the foci of converging lenses and mirrors; and also a shadow.

We can now define matter as *all that* which has both extension and impenetrability, and therefore void space or *vacuum* is space without impenetrability,—being penetrable.

3. *Figure.*

The bounding surfaces of a body give it its figure or form.

That department of science which discusses and classifies the forms of nature, is called Morphology.

The forms of bodies may be arranged under two divisions.

I. *Regular Forms.* II. *Irregular or Amorphous Forms.*

Regular forms can be embraced in two classes (*a* and *b*).

a. Forms of matter produced by the action of forces not directed by the vital principle.

1. Forms of the heavenly bodies. Considered in Astronomy.

Forms of liquids in motion; as "the liquid vein;" (see researches of Poncelet, Savart and Magnus), the rain drop; the forms of waves.

Forms of liquid at rest; (not contained in vessels) as the dew-drop; the soap-bubble.

Forms of liquids not subject to the action of gravity; as the forms assumed by oil—suspended in a mixture of water and alcohol of the same density as itself—when subjected to various conditions of molecular action. (See Plateau's *experimental and theoretical researches on the figures of equilibrium of a liquid mass withdrawn from the action of gravity.* Translated in Smithsonian Reports, 1863, *et seq.*

2. Forms of crystals. The laws ruling these forms considered in the science of Crystallography.

b. Forms of matter produced by the action of forces directed by the vital principle.

1. Forms of plants. Considered in botany.

2. Forms of animals. Considered in zoölogy.

All the forms of matter not contained in the foregoing classes are irregular or *amorphous*, and belong to class II.

Diagrams illustrating above classification of forms.

4. *Divisibility.*

Every body can be divided into many parts, and these parts can be further subdivided until the parts become so minute as to escape observation, even when aided by the most powerful microscopes.

That matter was susceptible of very minute division, was known to the philosophers of ancient Greece and of India, and they discussed the question, whether matter was infinitely divisible, or divisible only to a certain minuteness, when the parts were supposed to be unalterable by any means, whether mechanical or chemical. Mathematically speaking, the division has no limit, for however small the residual particles may be, nevertheless, since they have extension they have divisibility. But we demand not what may be conceived, but what really exists; and in fact, the discussions of the ancients are about the infinite divisibility of space, and prove nothing as to the actual divisibility of matter.

Anaxagoras and Aristotle admitted its infinite divisibility; Leucippus maintained a contrary opinion, and Democritus, siding with Leucippus, gave the name of *atoms* to the ultimate unalterable parts of which he supposed all bodies to be composed. Epicurus attempted to develope the ideas of Democritus, that found in the seventeenth century a zealous defender in Gassendi, whilst Descartes upheld the opinion of Aristotle.

But we have in the phenomena of molecular physics, in the laws of combination in chemistry, and in the facts of crystallography strong evidence that all bodies are composed of ultimate excessively small parts, which are invisible, even with the aid of the most powerful microscopes, but still of finite dimensions, of infinite hardness, unalterable by any means, and separated from each other by spaces which are very great when compared with their size. These parts which are the limits of the possible division of matter, are called *atoms* (Gr. ἄτομος; α privative and τέμνω, to cut.)

According to this hypothesis, a union or grouping of two or more atoms forms a *molecule;* a combination of molecules a *compound molecule*, and a union of the latter a particle.

The sensible division of matter can be carried to a very minute limit.

Examples. Gold can be beaten into leaves $\frac{1}{25000}$th of a millimètre thick.

Silver wire, gilded with $\frac{1}{360}$th of its diameter of gold, can be drawn so fine that one mètre weighs only eight milligrammes. The gold film on this wire is now only $\frac{1}{800000}$th of a millimètre thick. By placing a short piece of this wire in nitric acid, the silver core is dissolved out, leaving a tube of gold, having a wall only the $\frac{1}{800000}$th millimètre thick. Now, under the best microscopes, we can discern a surface of $\frac{1}{4000}$th of a millimètre in diameter; therefore we can divide gold into particles $\frac{1}{4000}$th millimètre in diameter, and $\frac{1}{800000}$th millimètre thick; yet each of these parts has all the physical and chemical properties of a large mass, as can be determined by testing it under a microscope.

Dr. Wollaston drew platinum wire so fine that its diameter was only $\frac{1}{1200}$th millimètre ($\frac{1}{30000}$th inch); and although platinum is the heaviest of the metals, yet it took 200 mètres of this wire to equal one centigramme in weight; or, in other words, one mile of this wire weighed about one grain. Dr. Wollaston accomplished this by wire-drawing a cylinder of silver $\frac{1}{5}$th of an inch in diameter, having in its axis a platinum wire $\frac{1}{100}$th inch diameter, and after having obtained a very fine wire—having in its interior a platinum wire of still greater tenuity—he dissolved with nitric acid the silver coating, and thus obtained an almost invisible platinum wire.

The thickness of a soap-bubble, in the dark spot which is formed on it just before it bursts, is $\frac{1}{100000}$th millimètre.

Divisibility of matter in solution. One grain of carmine tinges ten pounds of water, which we can divide into about 617,000 drops. If we suppose that 100 particles of carmine are requisite to produce a uniform tint in each drop, it follows that the one grain of carmine has been divided into 62,000,000 parts.

Metallic solutions and chemical tests.

Illustrations from organized bodies.

The thread of a spider is composed of more than 1000 separate threads.

The diameter of the red disks contained in human blood is $\frac{1}{4000}$th inch; while the diameter of the blood-disks of the Java musk-deer is only the $\frac{1}{12325}$ inch, so that a drop of this deer's blood, such as would adhere to the point of a fine needle, would contain 150,000,000 disks.

It has been calculated that some of the siliceous plates which cover and give rigidity to the minute vegetable cell-plants, called diatomaceæ, weigh only $\frac{1}{2000000000}$th of a grain; yet the surfaces of these plates are covered with the most exquisite tracery of siliceous stria or bars, often not more than $\frac{1}{85000}$th inch in width and thickness. Now we can discern a surface of $\frac{1}{85000}$th inch in the best microscopes, and a portion of one of these siliceous disks of that area would weigh only about $\frac{1}{4500000000}$th of a grain.

Divisibility of odorous substances.

A portion of musk will give off a powerful odor during a year, and yet its diminution in weight has not been sufficient to be detected by the most delicate balance.

"In order to offer an inexact idea of the minuteness of the particles of musk which are still capable of imparting some odor, we state, after a well known experimenting physiologist, that a certain liquid, containing as much of an extract of spirit of musk as $\frac{1}{2000000000}$th part of its whole weight, was at this time still distinctly odorous. A grain's weight of a liquid of which $\frac{1}{2000000}$th part was of that extract, spread an intensely penetrating odor. Next after musk are to be mentioned certain flower ethers, especially the oil of roses, a little drop of which is sufficient to fill with odor an immense atmosphere. The same physiologist states that a certain space filled with air, of which, at the highest, only $\frac{1}{1000000}$th part was vapor of oil of roses, still diffused a distinct odor of roses." See article "On the Senses," in Smithsonian Report, 1865.

5. *Compressibility.*

All bodies when subjected to exterior pressure are reduced in volume, and since all bodies can thus be indefinitely compressed, it follows that the matter of which bodies are composed does not fill the space which is contained within their bounding surfaces; or, as it is sometimes stated, a body does not form a *plenum* of matter.

The compressibility of solids is seen in all structures and machines where matter experiences great pressure; *e. g.* The stone pillars which support the dome of the Pantheon, at Paris, sensibly diminished in height as the weight which they bear was placed upon them, and a visible depression takes places in the arches of massive bridges when the "centres" are knocked away.

The relief of coins and medals is produced by subjecting disks of metal, placed between hard steel dies, to an intense and sudden pressure, which is so great that the metal disk is not only changed in form, but its *volume* after being struck is appreciably less than before.

For a long time, liquids were regarded as incompressible, and this opinion seemed to be confirmed by many experiments which the Academy of Florence made during the end of the seventeenth century. Their most celebrated experiment, which, however, had previously been made by Lord Bacon, consisted in filling a hollow silver sphere with water, and after stopping the orifice with a screw-plug, subjecting it to the powerful pressure of a screw-press. It is evident that if the sphere is flattened, its capacity is diminished, and the water compressed. They were surprised, however, to see the water appear on the exterior surface, as a fine dew, and they thence concluded that water was incompressible, and silver porous to this liquid.

John Canton, in Trans. R. S. 1762, first showed that liquids were compressible, even with the pressure of one atmosphere. His apparatus consisted of a large thermometer which contained the liquid, and the bulb of which was inclosed in an exhausted receiver. The liquid was thus entirely relieved of atmospheric pressure. The height of the liquid was now marked upon the stem, and the end of the sealed thermometer tube being broken, the air was allowed to enter the stem and press upon the surface of the liquid, and to enter the receiver and press upon the outside of the bulb. The liquid instantly fell in the tube, thus clearly showing a compression produced by the pressure of one atmosphere.

Canton thus found that water was compressed $\frac{46}{1000000}$ths of its volume by the pressure ef one atmosphere, and this determination is quite exact.

In 1826, Jacob Perkins described in the Trans. R. S. an appparatus which renders the compressibility of liquids very evident. Into a hollow cylinder passed a rod through a stuffing-box, and around this rod tightly fitted a leather washer, which rested on the top of the cylinder. This apparatus being filled with water, or other liquid, was placed in a closed cannon, into which water was forced by a strong pump, until a certain pressure was attained, which was measured by the lifting of a safety-valve and the escape of water.

On taking out the cylinder from the cannon, the leather washer was found on the rod several inches above the place it occupied before the experiment, thus showing that the rod had entered by that quantity into the vessel, as the pressure which it exerted compressed the liquid.

Describe and use Œrsted's piezometer, and exhibit table of Regnault and Grassi's results on the compressibility of liquids.

Gases, of all bodies, are the most easily compressed. This fact is readily shown by enclosing a gas in a cylinder, into which glides an air-tight piston. By pressing on the piston, it is forced into the cylinder, and the gas is reduced in volume. On relieving it of pressure, the gas expands to the volume it had before the experiment, and will do so no matter into how small a volume it may have been compressed, or how often the experiment may have been made. In this respect (of perfect elasticity) gases and liquids differ from solids. This is explained by the fact, that the ultimate parts of solids have *polarity*.

It is found that by pressure, gases are reduced to volumes, which are the reciprocals of the pressures.

6. *Dilatability.*

When any body is subjected to exterior pressure on all its surfaces, it is forced into a smaller volume, but on the pressure being removed, it expands to the dimension it had before the pressure was applied. Also, the volumes of all bodies are increased by a rise in their temperature; and Cagnard—Latour has shown that wires and rods when stretched, have their volumes increased.

This general.property of increase in volume from these causes is called dilatability.

The dilation of solids and of liquids is shown by their expansion, when relieved of hydrostatic and atmospheric pressures; and any gas expands when the atmospheric pressure is removed.

Experiments. The expansion of solids shown when relieved of powerful pressure. The experiment of Canton *reversed*, shows the expansion of a liquid as the pressure of the atmosphere is removed.

The expansion by heat, shown by Saxton's pyrometer, of liquids and of gases by large liquid and air thermometers.

Experiments of stretching wires and India-rubber, with Cagnard—Latour's apparatus.

7. *Porosity.*

The spaces which exist between the ultimate parts of bodies—that is, between their atoms and between their molecules—are called *pores;* by virtue of which all bodies have porosity.

This property is a natural deduction from the properties of compressibility and dilatability; for all bodies can be reduced in volume by pressure, and this reduction of volume is only limited by the pressures which our machines can produce, and all bodies expand in volume when relieved of pressure; we therefore conclude that the ultimate parts of bodies do not touch, but are separated by intervals that we call pores.

This property of porosity, first indicated and concisely defined by Gassendi, can, however, be directly proved by numerous experiments.

Porosity of solids. Water and oil have, by enormous pressures, been forced through the walls of vessels of gold and of iron; iron converted into steel by absorbing carbon at a high temperature; gold and the densest bodies are transparent when reduced to sufficiently thin leaves; lead is so porous to mercury, that Prof. Henry made a syphon of a solid bar of lead draw mercury over the side of a vessel containing it; marble and all rocks are porous to gases, allowing them to "diffuse" through them. We can ascertain how a stone will be affected by frost, by allowing it first to absorb a solution of sulphate of soda, and then ascertaining the effects produced by its subsequent crystallization. It is better, however, to determine directly the effect of the freezing of absorbed water. See *Smithsonian Report*, 1857, p. 305. (It is probable that the two preceding instances are examples of the porosity existing between the integral crystals of the stone and not between the ultimate parts, and therefore is not porosity in the sense in which we have defined it.) Hydrophane, a species of agate, becomes translucent by the absorption of water, when immersed in that liquid, while the displaced air issues from the hydrophane in minute bubbles.

M. M. Deville and Troost have recently shown that platinum and other dense metals and cast iron, when heated to redness, are porous to gases. The injury to health produced by heating dwellings by means of cast iron stoves, is caused by the vitiation of the atmosphere by the gases of combustion transmitted through the hot iron. (See *Comptes Rendus*, Jan. 20, 1868. Report of commission on the observations of Dr. Carret, on the deleterious effects of cast iron stoves.)

Porosity of Liquids. If measured volumes of water and of alcohol are mixed, it is found that their combined volume is less than the sum of their volumes before mixing. The greatest contraction takes place when 100 volumes of water are combined with 116 volumes of anhydrous alcohol, the diminution in volume being equal to 3·7 per cent.

This experiment, due to Réaumur, is made as follows: to a bottle is adapted a deeply fitting cork, perforated by a glass tube. The requisite proportion of water is placed in the bottle, and on this, colored alcohol is gently poured until it completely fills the bottle. The stopper is now forced in, and the alcohol rises in the glass tube. The height of the liquid is marked, and the bottle being so inclined that the mixture of the liquids takes place, the fluids contract, and the alcohol descends in the tube.

Contraction also takes place when water is mixed with sulphuric acid, with salt and with other substances. No experiment could more clearly show that the atoms of water and alcohol do not really fill the space which the liquids appear to occupy.

Water and other liquids dissolve gases; for example, 1 volume of water at 60° Fah. will dissolve 720 volumes of ammonia, and increase in its volume only one-half, and in its weight one-third.

Porosity of gases. Six parts by weight of carbon will unite with eight parts by weight of oxygen, and yet the resulting carbonic acid gas has the volume of the oxygen before combination.

Air and vapor are porous to each other, and between the atoms of air there may exist the ultimate parts of other gases.

It is to be remarked that this property of porosity is not opposed to the property of impenetrability, for it is the atoms which are impenetrable, while it is the intermolecular spaces of one body which are penetrated by the atoms of another.

We must be careful not to confound the true meaning of porosity, as given above, with what we designate as *organic*, or *structural porosity.* Organic pores are interstices which are visible with the aid of the microscope, and often even with the naked eye, while the intermolecular spaces are invisible by any means.

Examples of organic porosity. Mercury forced through wood. Circulation of sap in plants.

Examples of structural porosity. Porosity of stones, caused by the minute interstices produced by the crystalline structure of the stone. Paper used to filter liquids.

8. *Mobility.*

Everywhere in the universe, we see masses of matter changing their positions with reference to each other; while others appear to be at rest.

We only know that a body changes its position by referring it to other bodies, supposed to be at rest; and therefore Descartes defined motion as a rectilinear change of distance between two points. From this definition, if only one point existed in the universe, its conditions, whether of rest or of motion, could not be determined.

If the points to which we refer the moving body be really at rest, then the successive varying measures give us the *absolute motion*, but if this point be only apparently at rest, we obtain the *relative motion.* Give illustrations.

Rest is however only *apparent;* we know of no point in the universe which is absolutely at rest; therefore all the motions which we have determined are relative.

All bodies on the surface of the earth have the relative positions of their atoms changed by every change of temperature, by every vibration which they transmit; they are in motion with the earth on its axis and in its orbit, while the earth with the planets and sun are being translated in an unknown path in space. The stars, improperly called *fixed*, have also motions of their own, which, in some cases, are evidently of great magnitude.

Kind of motions. A motion may be one of *translation*, of *rotation* round an axis, or of a combination of these motions; it may be *rectilinear* or *curvilinear*, *continued* or *reciprocating.* Give illustrations.

The line a point of a body describes in space, is called its *trajectory.*

The *velocity* of motion of a moving body, or the ratio of the space described to the time of describing it, may be *uniform*, *uniformly accelerated*, *uniformly retarded*, or *irregular.*

Uniform motion is that in which equal spaces are passed over in equal, small portions of time.

Formulæ of uniform motion.

$$(1)\quad S = V T;\ T = \frac{S}{V};\ V = \frac{S}{T}$$

in which S = space described; V = velocity of moving body, or the space described in unit of time; T = time occupied in going over space S.

Uniformly varied motion is that in which the change in velocity at the end of a certain time, is proportional to this time.

Let u be the initial velocity of the body, that is to say, the velocity at a given instant from which we commence to reckon the time; a the acceleration, or the constant quantity by which the velocity varies in a unit of time; v the velocity at the end of t seconds; we have from definition of uniformly varied motion.

$$(2) \quad v = u \pm a\,t$$

The sign + when the motion is accelerated; the sign — when it is retarded. In the last case, the velocity will become o when $u = a\,t$, that is to say, at the end of a number of seconds represented by $u : a$.

If, in formula (2), we make $u = o$, then the moving body starts from a state of rest, and we have at the end of time t

$$(3) \quad v = a\,t;$$

that is, *the velocity acquired at the end of a certain time is proportional to this time.*

In uniformly accelerated motion, the spaces described by a body starting from a state of rest, are proportional to the squares of the times employed to describe them; or—

$$(4) \quad s = \tfrac{1}{2}\,a\,t^2$$

which shows that the space gone over in uniformly accelerated motion, by a body which starts from a state of rest, is equal to the space which it would go over with a *uniform motion* with the velocity $\frac{1}{2}\,v$ or $\frac{1}{2}\,a\,t$.

It is often required to know the velocity acquired in function of the space gone over. We obtain this by eliminating t in the two equations $v = a\,t$, $s = \frac{1}{2}\,a\,t^2$, which gives—

$$(5) \quad v = \sqrt{2as}$$

A body falling vertically, in vacuo, by the action of gravity, from a height which is exceedingly small compared with the radius of the earth, may be regarded as having a uniformly accelerated motion. In this case a is equal to the velocity acquired by the body at the end of the first second of its fall. This velocity which is the measure of the intensity of gravity, and which is equal, in the latitude of New York, to 32 feet 2 inches, is always indicated by the letter g. The formula (5) then becomes—

$$v = \sqrt{2\,g\,s}$$

Problem. What velocity will a body acquire by falling 772 feet, in the latitude of New York? Ans. 222 feet, 10·28 inches.

When the moving body possesses an initial velocity u, at the moment from which we count the time, the above formulæ become—

$$v = u \pm a\,t;\; s = u\,t + \tfrac{1}{2}\,a\,t^2,\; v = \sqrt{u^2 + 2\,a\,s}$$

To Galileo is due the discovery of the laws of uniformly varied motion. Give his classical demonstration of these laws.

Co-existence of separate motions. When a body in motion is acted on by a force, the same effect in motion is produced as if that force moved the body from a state of rest. Thus, a cannon ball projected horizontally from an elevation, will reach the ground by the action of gravity, in the same time (assuming that no air resists its motion) as if it dropped vertically from the elevation through the same height.

"A body describes the diagonal of a parallelogram by two forces acting conjointly, in the same time in which it would describe its sides by the same forces acting separately."—*Newton.*

The discovery of the law of the co-existence of motions, is also generally ascribed to Galileo (Dial. 4, Prop. 2), but Aristotle clearly announced it in his Mechanical Problems, c. 24.

The various relations of space and time can readily be represented geometrically, by making lengths of ordinates stand for velocities acquired at the end of times represented by lengths of abscissas. Thus, uniformly varied motion is represented by a right angled triangle, whose base equals in units of length the units of time during which the motion took place, its altitude, the velocity acquired at the end of this time, while the units of area of the triangle represents the distance gone through by the moving body.

The consideration of motions, irrespective of force and of the properties of matter, constitutes the part of science called *Kinematics.* (Gr. Κινησις, motion.) The ablest discussion of this subject is found in first chapter of the *Natural Philosophy*, by Profs. Thomson and Tait, Oxford, 1867.

9. *Inertia—Force.*

The property of matter by which it tends to retain its state, whether of rest or of motion, is called inertia.

By saying that a body has inertia, we merely understand that a body cannot *of itself* modify its condition, whether of rest or of motion; and that whenever a body begins to move, or to change the velocity or the direction of its motion, these changes in its condition are to be referred to some extraneous cause.

When a body is set in motion and abandoned entirely to itself—when it is conceived as being alone in space—it will move in *a straight line*, which is the direction of its first motion, and with its first velocity forever. This truth, called the law of inertia, is the result of an extended induction, and was not recognized before the time of Kepler. Descartes made it the foundation of his principles of mechanics.

Give illustrations of above principle, from observations of the motions of the heavenly bodies, and from experiments on the motions of bodies on the surface of the earth. The rotation of the earth on its axis. A pendulum will vibrate two days in a vacuum, when the friction of the point of suspension is reduced to its minimum.

The apparent departures from the law of inertia, can all be referred to the action of forces or of resistances exterior to the body, and which are opposed to its uniform motion in a straight line.

Force.

All the phenomena, or changes which we observe in the condition of matter, are motions or the results of motions of either masses or of their ultimate parts or atoms; and that which produces these changes in the condition of matter is denominated *force.*

To Dr. Julius Robert Mayer, of Heilbronn, Germany, we owe the first successful attempt to give as clear conceptions in reference to force, as previously existed in relation to matter. In 1842, he published in *Liebig's Annalen*, a short paper of eight pages, entitled *Bemerkungen über die Kräfte der unbelebeten Natur*, which from the fundamental importance of the truths which it unfolds, and from the results which have been deduced from them, is to be considered as one of the most important additions to knowledge produced in this century.

Mayer reasons thus: "Forces are causes; accordingly, we may in relation to them, make full application of the principle, *causa, æquat effectum.* If the cause c has the effect e, then $c = e$; if, in its turn, e is the cause of a second effect, f, we have $e = f$, and so on; $c = e = f \ldots = c$. In a chain of causes and effects, a term or a part of a term can never, as plainly appears from the nature of an equation, become equal to nothing. This first property of all causes we call their *indestructibility.*

"If the given cause, c, has produced an effect, e, equal to itself, it has in that very act ceased to be; c has become e; if, after the production of e, c still remained in whole or in part, there must be still further effects corresponding to this remaining cause; the total effect of c would thus be $>e$, which would be contrary to the supposition $c=e$. Accordingly, since c becomes e, and e becomes f, &c., we must regard these various magnitudes as different forms under which one and the same object makes its appearance. This capability of assuming various forms, is the second essential property of all causes. Taking both properties together, we may say causes are (quantitatively) *indestructible* and (qualitatively) *convertible* objects."

In another important paper, "*Bemerkungen über das mechanische Aequivalent der Wärme*, 1851," Mayer says: "*Force is something which is expended in producing motion;* and this something which is expended is to be looked upon as a cause equivalent to the effect, namely, to the motion produced."

Now, in these motions or effects, there are evidently two things to be considered, (1) the *mass* of matter moved, and (2) the *space* through which it is moved; and we have therefore force $=$ mass $\times$ space gone through; but as we can *measure* and *compare* forces only by measuring and comparing their effects, and as bodies in free motion will move in the same right line, and with uniform velocity forever, we must place the moving bodies in such circumstances that their motions are destroyed, and we have remaining in their stead, equivalent effects, which we can *measure;* then the comparison of these measures will give us the relative intensities of the forces.

These effects, either directly or indirectly obtained from the moving body, are as various as there are kind of forces and resistances existing; thus, we may oppose to a body, moving vertically upward, the resistance of gravity (which we may regard as constant, if the upward flight of the body is a very minute fraction of the radius of the earth); or, we may oppose the constant resistance, which a body of homogeneous structure offers when it is penetrated by another, as, for example, when a cannon ball penetrates earth, clay, or pine wood; or, again, we may have, for the effect of the destroyed motion, *heat*, which makes its appearance whenever a moving body is brought to rest either by friction, percussion, compression, &c., with some other body, or, as in Foucault's experiment, where a copper disk being forced to revolve between the poles of a powerful electro-magnet, the motion of the wheel being opposed by the reaction existing between the electric currents flowing in it and in the magnet, the motion (force) lost by the wheel appears as heat (force) in its substance.

The heat which is produced by any of these means can readily be caused to evolve, as *it* disappears, dynamic electricity, light, and chemical action. Thus, in an experiment which the author devised for his classes, about four years ago, the *heat* developed by the "*falling force*" of a weight striking the terminals of a compound thermal-battery, formed of pieces of iron wire and German silver wire twisted together at alternate ends, caused a current of electricity through the wires, which being conducted through a helix *magnetized* a a needle (which then attracted fine iron particles), caused *light* to appear in a portion of the circuit formed of Wollaston's fine wire, *decomposed* iodide of potassium, and finally *moved* the needles of a galvanometer.

Let us now try to arrive at comparable measures of force by first opposing to the moving body the constant resistance of gravity, and see if the measures thus given for different velocities compare with measures given by other resistances overcome, and for different quantities of heat, electricity, &c., developed by the disappearance of different velocities in the moving mass.

A body, shot vertically upward, with a velocity v, comes to rest, by the opposing resistance of gravity, after having reached a certain height, which we will call h. Giving the body twice the initial velocity, $2v$, it reaches $4h$ before it begins to return to the earth; with the initial velocity $3v$, it reaches the height $9h$; while $4v$ gives $16h$, and so on; in other words, *the heights reached are as the squares of the initial velocities.*

Wherefore, as force $=$ mass $\times$ space gone through, it follows that the measure of force is *mass* $\times v^2$; v being the velocity of the mass; or, *force* $=$ *mass* $\times$ *distance gone through* in overcoming the constant resistance $=$ *mass* $\times v^2$.

If this measure be true, the same ratio of the square of the velocity, will exist when other resistances are opposed to the moving body. Take the resistance offered by an earth or clay bank to the penetration of cannon balls having different velocities. It is found by artillerists, that a ball striking with the velocity $2v$, will penetrate four times as deep as the same ball with velocity v; while a velocity of $3v$ will give a penetration of nine times the depth and so on; the penetration of the same ball being as the squares of its velocities. (See Dr. Wollaston's Bakerian Lecture on the Force of Percussion, Phil. Trans. 1806; and Benton's Ordnance and Gunnery, N. Y., 1862, p. 476, *et seq.*)

Having found this measure of force true for these two cases, where motion disappears, let us determine, as we only can, *experimentally* and *inductively*, whether the relative quantities of *heat* evolved as different velocities of the moving mass disappear, also preserve the ratio of the squares of those velocities.

In the year 1850, there appeared in the Phil. Trans. R. S. Lond., a paper "*On the Mechanical Equivalent of Heat,*" by Dr. James Prescott Joule, of Manchester, England. This memoir contains one of the most important physical constants ever determined.

In this investigation was first obtained, *by direct experiment*, the exact quantity of heat developed by the falling of a given weight through a known height. His experiments on this subject began in 1843, and were continued during six years, until 1849. Dr. Joule, during this long experimental experience, gradually perfected his apparatus, and learned to eliminate various sources of error, until finally his measures arrived at by different processes gave, within small limits, almost identical values for "the mechanical equivalent of heat."*

His apparatus consisted of an upright copper cylinder, which contained either water, oil, or mercury; in the lid of this vessel were two apertures, one for a vertical axis, to revolve in without touching the lid, the other for the introduction of a thermometer. The vertical axis, which was perfectly fitted into the bottom of the vessel, carried eight revolving arms or paddles, which, as they went round, passed between openings in four stationary vanes, so that the water could not acquire a motion of rotation and move with the arms; and resistance was thus made to their motion. Two weights were attached to fine flexible cords, which passed over pulleys, and were wound round a roller on the vertical axis, armed with the paddles. These weights in falling caused the paddles to revolve, and by the resistance which it opposed to their motion, the liquid was heated by the equivalent in motion expended. The height of

* It is to be remarked that Mayer, in 1842, used the expression "mechanical equivalent of heat," and from the difference in the specific heats of the same weight of air under constant pressure, and under constant volume, theoretically *deduced* a value for this equivalent, which, corrected with the exact data of the above quantities as furnished by Regnault, gives a result nearly identical with Joule's.

the fall was about sixty-three feet (to which we may say that practically the radius of the earth was infinite), and was measured by vertical scales, along which the weights descended.

The mode of experimenting was as follows: the temperature of the liquid having been ascertained by a thermometer, which was capable of indicating a variation of temperature as small as $\frac{1}{200}$th of a degree Fah., and the weights wound up by detaching the roller from the vertical paddle-axis, the precise height of the weights were ascertained after keying the roller; the weights then descended until they reached the floor. The roller was again detached from the paddle-axis, the weights wound up, and the agitation of the liquid renewed. This was repeated twenty times, and then the temperature of the liquid was again observed. The mean temperature of the room was derived from observations made at the beginning, middle, and end of each experiment.

Corrections were now made for the effects of radiation and conduction; and, in the experiments with water, for the quantities of heat absorbed by the copper vessel and by the paddle wheel. In the experiments on the heat produced by the agitation of mercury, and in the heat given out by the rubbing of cast iron plates, the heat capacity of the entire apparatus was ascertained by observing the heating which it produced on a known weight of water in which it was immersed. In all the experiments, corrections were also made for the velocity with which the weights came to the ground, and for the rigidity of the strings. The force expended in friction of the apparatus was diminished as far as possible by the use of friction wheels, and its amount was determined by connecting the pulleys without connection with the paddle-axis, and ascertaining the weight necessary to give them a uniform motion.

The following table gives the results of Joule; the second column, as they were obtained in air, the third, the same corrected, as though the weights had descended in vacuo.

MATERIALS EMPLOYED.	MEAN EQUIV. IN AIR.	MEAN EQUIV. IN VAC.	NO. OF EXP'S FROM WHICH DERIVED.
Water	773·640	772·692	40
Mercury	775·032	774·083	50
Cast Iron	775·938	774·987	20

In the experiments of producing heat by the rotation of one cast iron plate on another, the friction produced considerable vibration in the frame work of the apparatus, and a loud sound; allowance was therefore made for the quantity of force expended in producing these effects.

The number 772·692, obtained as the mean of forty experiments on the friction of water, Joule considered the most trustworthy; but this he reduced to 772, because, even in the friction of liquids, he found it impossible entirely to avoid vibration and sound. The deductions of Joule from these experiments are :

1. *That the quantity of heat produced by the friction of bodies, whether solid or liquid, is always proportional to the force expended.*

2. *That the quantity of heat capable of increasing the temperature of one pound of water (weighed in vacuo, and between* 55° *and* 60°) *by* 1° *Fah., requires for its evolution the expenditure of a mechanical force represented by the fall of* 1 *pound through* 772 *feet, or* 772 *foot-pounds.*

This is the "*Mechanical Equivalent of Heat,*" or the unit of heat, generally called in honor of the illustrious physicist, "Joule's Unit."

In French measures, the above heat-unit is thus stated: *The heat capable of increasing the temperature of* 1 *kilogramme of water* 1° *C., is equivalent to a force represented by the fall of* 423·55 *kilogrammes, through the space of* 1 *mètre.* The descent or ascent of 1 pound through 1 foot is called *a foot-pound*, while the descent or ascent of 1 kilogramme through 1 mètre is denominated a *kilogramme-mètre.* By the adoption of these terms, the expressions of the above truths can be mere concisely enunciated; thus, using French measures, we say, 423 *kilogramme-mètres is equivalent to* 1 *kilogramme-degree centigrade.*

Thus, Joule showed that the heat developed was in proportion to the mass × the distance fallen through; or, what is the same, equivalent to the mass × square of the velocity.

In a remarkably interesting paper, "*On the Production of Thermo-Electric Currents by Percussion,*" Prof. O. N. Rood, of Columbia College, N. Y., shows directly by careful and skilful experiments, that the heat and its equivalent in dynamic electricity (which latter gave the measure of the heat), produced by a weight falling from different heights on a compound plate of German silver and iron, was in proportion to the height of the fall of the weight, or, what is the same, to the square of the velocity of impact.

For the details of these experiments, and the precautions taken to avoid the action of extraneous causes mingling themselves with the main effect, the reader is referred to Prof Rood's paper in "The American Journal of Science, July, 1866."

We here only give some of the results, which speak for themselves.

TABLE 2.—WITH TWO SKINS ABOVE AND BELOW JUNCTION OF METALS.				
Distances fallen by Weight	1 in.	2 ins.	3 ins.	4 ins.
Deflection of Galvanometer-needles —Average of eight Observations	1°·3	2°·7	3°·55	5·°2

TABLE 3.—WITH FOUR LAYERS OF PLAIN SILK ABOVE AND BELOW THE JUNCTURE.				
Distances Fallen	1 in.	2 ins.	3 ins.	4 ins.
Average Deviation of eight exp'ts.	1°·5	3°·0	4°·5	6°·0

TABLE 4.—WITH FOUR LAYERS OF WAX-COATED WOVEN SILK.					
Distances Fallen	1 in.	2 ins.	3 ins.	4 ins.	5 ins.
Reduced Average Deviation of Five Experiments	1°·07	2°·1	3°·28	4°·3	6°·0

For a proper interpretation of the above results, it is to be remarked that the deflection of the galvanometer needles up to 6° was in proportion to the intensity of the current, which is itself, within these limits, proportional to the heat at the juncture which developed it.

It can be further shown that the measure of force, or *energy* (a term now more generally used, and which we owe to Dr. Thomas Young), is proportional to the square of the velocity of a moving mass, by obtaining the equivalents of effects other than those of resistance offered by gravity, by penetrable bodies, and of the heat developed by friction and by impact. Joule, in 1843, showed that the same relation existed between the heat evolved by the electric current of an electro-magnetic engine, and the mechanical energy expended in producing it, and in 1844 he showed that the heat absorbed and evolved by the rarefaction and condensation of air, is proportional to the amount of mechanical energy evolved and absorbed in these operations.

In the following table, taken from Verdet's "*Exposé de la Théorie Méchanique de la Chaleur*," *Paris*, 1863, are given the most reliable determinations of the mechanical equivalent of heat. The numbers represent the number of kilogramme-mêtres, which is equivalent to one kilogramme-degree centigrade of water.

NATURE OF THE PHENOMENON WHENCE THE DETERMINATION IS DRAWN.	PHILOSOPHERS WHO HAVE INVENTED THE THEORETICAL PRINCIPLES OF THE DETERMINATION.	PHILOSOPHERS WHO HAVE SUPPLIED THE EXPERIMENTAL DATA.	MECHANICAL EQUIVALENT.
General Properties of Air........	Mayer, Clausius	V. Regnault, Moll & Van Beek	426
		Joule	424
Friction............................	Joule	Favre	413
Work done by the Steam Engine	Clausius	Hirn	413
Heat evolved by Induced Currents	Joule	Joule	452
Heat evolved by an Electro-Magnetic Engine at Rest and in Motion.........................	Favre	Favre	443
Total Heat evolved in the circuit of a Daniell's Battery...	Bosscha	W. Weber, Joule	420
Heat evolved in a metallic wire, through which an electric current is passing..............	Clausius	Quintus Icilius	400

From the above discussion, we conclude that the true measure of force is *mass* $\times v^2$.

The Theory of Energy considers its Conservation, its Transformation and its Dissipation.

The *Conservation of Energy.*—The total amount of energy in the universe, or in any limited system which does not receive energy from without, or part with it to external matter, is invariable. Energy is, in other words, as indestructible as matter, and is neither created nor destroyed, but merely changes its form.

The *Transformation of Energy.*—By an extended induction, we find that any one form of energy may be transformed, wholly or partially, to an equivalent amount in another form. These transformations are, however, subject to limitations contained in the principle of

The *Dissipation of Energy.*—"No known natural process is exactly reversible, and whenever an attempt is made to transform or retransform energy by an imperfect process, part of the energy is necessarily transformed into heat and dissipated, so as to be incapable of further useful transformation. It therefore follows that as energy is constantly in a state of transformation, there is a constant degradation of energy to the final unavailable form of uniformly diffused heat; and this will go on as long as transformations occur, until the whole energy of the universe has taken this final form." See *N. Brit. Rev.*, May, 1864.

"There is consequently," says Prof. Thomson, "so far as we understand the present condition of the universe, a tendency towards a state in which all physical energy will be in the state of heat, and that heat so diffused that all matter will be at the same temperature; so that there will be an end of all physical phenomena."

Vast as this speculation may seem, it appears to be soundly based on experimental data, and to represent truly the present condition of the universe, so far as we know it. See Prof. Thomson "*On a Universal Tendency in Nature to the Dissipation of Mechanical Energy.*"—Proc. R. S. Edinb. and Phil. Mag., 1852.

The energy of a *moving* body is the work which it is capable of performing against a resistance before being brought to rest, and is equal to the force which must act on the body to move it from a state of rest to the given velocity. This force is measured by the product of the mass of the body into the height from which it must fall in order to acquire the given velocity; which is expressed thus:

$$\text{Energy} = \frac{M V^2}{2 g}$$

g representing as usual, the velocity acquired by a body at the end of the first second of its fall.

Energy may be of two kinds (1), *Kinetic energy*, or energy of motion, and (2) *Potential energy*, or energy of position or of condition. Thus, the energy of a ball shot vertically upward, is entirely kinetic at the moment of its discharge, while its energy is all potential, or one of position, when it has reached the summit of its flight and begins to descend. It is evident that the ball in descending, will gradually lose potential energy and gain kinetic energy (the sum of the two energies always remaining constant), and when it has reached the level from which it was discharged upward, its energy is again all kinetic, and equal to what it was when it began its upward flight. An oscillating pendulum is an instance where the energy is alternately kinetic and potential.

All the various forms of energy may be brought under two classes.

I. *Visible Energy*, or energy of visible motions and positions.
II. *Molecular Energy.*

Under class I. we have

A. Visible kinetic energy.
B. Potential energy of visible arrangement, as for examples, a head of water; a coiled spring; a raised weight.

Under class II.

C. The energy of electricity in motion.
D. The energy of radiant heat and light.
E. The kinetic energy of absorbed heat.
F. Molecular potential energy.
G. Potential energy caused by electrical separation.
H. Potential energy caused by chemical separation.

Now with regard to these various forms of energy, the principle of the conservation of energy asserts that for a body left to itself, or for the entire material universe, we must have

$$A+B+C+D+\&c.,=\text{a constant quantity.}$$

On the other hand, the various terms of the left hand member of this equation must be considered as variable quantities, subject, however, to the above limitation, but capable of being transformed into one another according to certain laws.

Laws of the transmutations of energy.—The following are among the most important cases of transmutation of these energies into one another.

A into B, when a weight is projected upwards; into C, when a conductor revolves between the poles of a magnet. A is not transmuted directly into D; it is into E, F and G. A is not directly converted into H.

B can be converted into A, and through it into other forms of energy.

C can be transmuted into A, into E, into F, and into H.

D can be transmuted into E, into F, and into H.

E and F is converted into A and into B, in the action of any heat-engine; into C, into D; into G when tourmalines are heated; and into H.

G can be converted into A and into C.

H can be transmuted into C; into E, into F, and into G. (See Elementary Treatise on Heat, by Balfour Stewart, Oxford, 1866.)

Sources of Energy.—The energy available for the production of mechanical work is almost entirely potential, and consists of

Potential forms of Energy.

1. Energy of Fuel.
2. Energy of Food.
3. Energy of Ordinary Water-Power.
4. Energy of Tidal Water-Power.
5. Energy of Chemical separation implied in native sulphur, native metals, free oxygen, &c.

Of Kinetic forms of Energy.

6. Energy of Winds and Ocean Currents.
7. Energy of Direct Rays of the Sun.
8. Energy of Volcanoes, Hot Springs and Internal Heat of the Earth.

"The immediate sources of these supplies of energy are four:—

I. Primordial Potential Energy of Chemical Affinity, which probably still exists in native metals, native sulphur, &c., but whose amount, at all events near the *surface* of the earth, is now very small.

II. Solar Radiation. III. The Earth's rotation about its axis.

IV. The Internal Heat of the Earth.

Thus, as regards (1) our supplies of fuel for heat-engines are, as was long ago remarked by Herschel and Stephenson, mainly due to solar radiation. Our coal is merely the result of transformation in vegetables, of solar energy into potential energy of chemical affinity. So, on a small scale, are diamond, amber and other combustible products of primeval vegetation. As Prof. Thomson remarks, wood fires give us heat and light which have been got from the sun a few years ago. Our coal fires and gas lamps bring out, for our present comfort, heat and light of a primeval sun, which have lain dormant as a potential energy, beneath seas and mountains for countless ages.

Though (II) thus accounts for the greater part of our store of energy, (I) must also be admitted, though to a very subordinate place.

As to (2), the food of all animals is vegetable or animal, and therefore ultimately vegetable. This energy, then, depends almost entirely on (II). This, also, was stated long ago by Herschel.

Ordinary water-power (3) is the result of evaporation, the diffusion and convection of vapor, and its subsequent condensation at a higher level. It also is mainly due to (II).

Tidal water-power (4), although not yet much used, is capable, if properly applied, of giving valuable supplies of energy. As the water is lifted by the attraction of the sun and moon, it may be secured by proper contrivances at its higher level, and then becomes an available supply of energy when the tide has fallen again. Any such supply is, however, abstracted from the energy of the earth's rotation (III). This was recognized by Kant; Mayer also and J. Thomson showed that the ebb and flow of the tides being due to the earth's revolving on her axis under the moon's attraction, the energy of the tides is really taken from the energy of the earth's revolution; part of which is thus ultimately dissipated in the heat of friction caused by the tides. The general tendency of tides on the surface of a planet is to retard its rotation till it turns always the same face to the tide-producing body; and it is probable that the remarkable fact that satellites generally turn the same face to their primary, is to be accounted for by tides produced by the primary in the satellite while it was yet in a molten state.

Winds and ocean currents (6), both employed in navigation, and the former in driving machinery, are, like (3), direct transformations of solar radiation (II).

As to (8), which is due to (IV), no application to useful mechanical purposes has yet been attempted.

We must next very briefly consider the origin of these causes, with the exception of (I), which is of course primary. Laplace, Mayer and Helmholtz come to our assistance, and suggest as the initial form of the energy of the universe, the potential energy of gravitation of matter irregularly diffused through infinite space. By simple calculations it is easy to see that, if the matter in the solar system had been originally spread through a space enclosing the orbit of Neptune, the falling together of its parts into separate agglomerations, such as the sun and planets, would far more than account for all the energy they now possess in the forms of heat and orbital and axial revolutions.

The sun still retains so much potential energy among its parts, that the mere contraction by cooling must be sufficient (on account of the diminution of potential energy) to maintain the present rate of radiation for ages to come. Moreover, the capacity of the sun's mass for heat, on account especially, of the enormous pressure to which it is exposed, is so great that (at the least and most favorable assumption) from 7,000 to 8,000 years must elapse, at the present rate of expenditure, before the temperature of the whole is lowered one degree centigrade, although the amount of solar heat received by the earth in one year is so enormous that it would liquify a layer of ice 100 feet thick, covering the whole surface of the earth, and if we bear in mind that the solar heat which reaches the earth in any time is only $\frac{1}{2300000000}$ of the heat which leaves the sun, we may obtain some idea of the immense heating power of the radiation from our luminary.*

It thus appears that if we except tidal-power, the sun's rays are the ultimate source of the available forms of energy with which we are surrounded.†

We see, from the above exposition, that "Philosophers have extended their ideas of quantity from matter to energy, and thus has arisen the new science of *Energetics*, or the quantitative study of the transformations of energy (as chemistry is the quantitative study of the transformations of matter), comprehending and uniting all the different branches of physical science."

* If the entire solar radiation were employed in dissolving a layer of ice, enclosing the sun, it would dissolve a stratum 10½ miles thick in a day.

† The sketch we here give of the *Sources of Energy*, is taken almost entirely from an article on "Energy,' in the N. Brit. Rev., May, 1864.

Efficiency of Heat-Engines.—After Joule had determined the mechanical equivalent of heat, engineers had the means of testing the actual efficiency of heat-engines.

If the number of thermal units produced by the combustion of one pound of a given kind of fuel, be multiplied by Joule's unit, 772 foot-pounds, the result is the *total heat of combustion* of the given fuel expressed in foot-pounds. This quantity ranges between 5,000,000 and 12,000,000 foot-pounds. But in the best existing steam-engines, it is found that on an average only about $\frac{1}{8}$ of the mechanical value of the heat produced by the fuel burning in the furnace, is obtained as useful mechanical effect, the remaining $\frac{7}{8}$ being wholly lost.

To understand the cause of this great loss, it is to be remembered that in every heat-engine the heat of the expansible fluid—which is the medium by which the heat of the fuel is transformed into the motion of the engine—disappears *as heat* by the exact equivalent, expressed in Joule's units, of the motion produced. Therefore the greater the fall in heat in the vapor, which, in expanding, cools and gives up its heat as motion, so will be the efficacy of the engine. Just as in a head of water, where the greater the difference between the higher and lower level, the greater the power obtained. But in the steam-engine we are obliged to obtain our power from the fall in temperature of the steam, which takes place in its expansion, so that in a steam-engine the power obtained is measured by the difference of temperature between the boiler and condenser, and *not* by the difference between the temperature of the furnace and condenser. Now, in the furnace the temperature is about 3,000 degrees above that of the atmosphere, while the temperature of the boiler is only about 200 degrees in excess of that of the condenser; therefore it is evident that the larger fall in temperature taking place between the furnace and the steam, the heat is lost or at least not utilized.

In a *perfect engine* the steam would enter the cylinder at the temperature of the furnace, and expand down until it had given up all its heat as motion to the piston, and would then enter the condenser at the temperature of the atmosphere; indeed, such an engine would require no condenser, for the steam would condense itself as the heat disappeared in its transmutation into motion.

We will conclude this sketch on the subject of Energy, with a concise statement in reference to the efficiency of heat-engines taken from Prof. Rankine; referring the reader who desires further information on this important and interesting subject to the works given below.

"The total heat produced in the furnace is expended, in any given engine, in producing the following effects, whose sum is equal to the heat so expended:—

1. The *waste heat of the furnace*, being from 0·1 to 0·6 of the total heat, according to the construction of the furnace and the skill with which the combustion is regulated.

2. The *necessarily-rejected heat of the engine*, being the excess of the whole heat communicated to the working fluid by each pound of fuel burned, above the portion of that heat which permanently disappears, being replaced by mechanical energy.

3. The *heat wasted by the engine*, whether by conduction or by non-fulfilment of the conditions of maximum efficiency.

4. The *useless work of the engine*, employed in overcoming friction and other prejudicial resistances.

5. The *useful work*. The efficiency of a heat-engine is improved by diminishing as far as possible, the first four of those effects, so as to increase the fifth.

"It appears, then, that the efficiency of a heat-engine is the product of three factors, viz: I. The *efficiency of the furnace*, being the ratio which the heat transferred to the working fluid bears to the total heat of combustion; II. The *efficiency of the fluid*, being the fraction of the heat received by it, which is transformed into mechanical energy; and III., The *efficiency of the mechanism*, being the fraction of that energy which is available for driving machines."

From the above discussion, we see immediately that by superheating the steam before it reaches the cylinder, we obtain a greater range of temperature for the steam to fall through in expanding, and thus render efficacious yet more of the heat of the furnace.

List of Works on the Conservation of Force and Thermodynamics.

The Correlation and Conservation of Forces; a collection of the papers of *Mayer*, *Helmholtz*, *Faraday*, *Grove*, *Liebig* and *Carpenter*. Edited by Dr. Edward L. Youmans, N. Y., 1865.

Joule.—On the Calorific Effects of Magneto-Electricity, and on the Mechanical Value of Heat. Phil. Mag., Vol. XXIII., 1843.

On the Changes of Temperature produced by the Rarefaction and Condensation of Air. Phil. Mag., May, 1845.

On the Mechanical Equivalent of Heat. Phil. Trans., 1850.

On some Thermodynamic Properties of Solids. Phil. Trans., 1859.

On the Thermal Effects of Compressing Fluids. Phil. Trans., 1859.

Clausius.—The Mechanical Theory of Heat, with its applications to the Steam Engine and to the Physical Properties of Bodies. Edited by T. A. Hirst, with an Introduction by Prof. Tyndall. London, 1867.

Thomson (William).—An Account of Carnot's Theory of the Motive Power of Heat. Trans. R. S., Edinb., 1849.

On the Dynamical Theory of Heat. Trans. R. S., Edinb., 1852.

Thomson and *Joule.*—On the Thermal Effects of Fluids in Motion. Phil. Trans., 1853.

On the Changes of Temperature experienced by Bodies moving through Air. Phil. Trans., 1860.

Rankine.—The Steam Engine and other Prime Movers. London, 1859.

Verdet.—Exposé de la Théorie Méchanique de la Chaleur. Paris, 1863.

Hirn.—Exposition Analytique et Expérimentale de la Théorie Méchanique de la Chaleur. Paris, 1865.

Saint—Robert.—Principes de Thermodynamique. Turin, 1865.

Bacon.—Novum Organum, De Formâ Calidi, book 2, aph. 20.

"Now from this our first vintage it follows, that the form or true definition of heat (considered relatively to the universe and not to the sense) is briefly thus:—Heat is a motion, expansive, restrained, and acting in its strife upon the smaller particles of bodies. But the expansion is thus modified: while it expands all ways, it has at the same time an inclination upwards. And the struggle in the particles is modified also; it is not sluggish, but hurried and with violence."

Locke.—"Heat is a very brisk agitation of the insensible parts of the object, which produce in us that sensation from whence we denominate the object hot; so what in our sensation is *heat*, in the object is nothing but *motion.*"

Rumford.—Trans. R. S. Lond., 1798. Rumford placed a cannon in a water-tight box, so that it could rotate against a blunt borer firmly pressed against the bottom of its chamber. The box was filled with water of a temperature of 60° F., and the cannon set in rotation by the power of horses.

"The result of this beautiful experiment was very striking, and the pleasure it afforded me amply repaid me for all the trouble I had had in contriving and arranging the complicated machinery used in making it. The cylinder had been in motion but a short time, when I perceived, by putting my hand into the water, and touching the outside of the cylinder, that heat was generated.

"At the end of an hour the fluid, which weighed 18·77 lbs., or 2½ gallons, had its temperature raised 47 degrees, being now 107 degrees.

"In thirty minutes more, or one hour and thirty minutes after the machinery had been set in motion, the heat of the water was 142 degrees.

"At the end of two hours from the beginning, the temperature was 178 degrees.

"At two hours and twenty minutes it was 200 degrees, and at two hours and thirty minutes it *actually boiled.*

"From the results of my computations, it appears that the quantity of heat produced equably, or in a continuous stream, if I may use the expression, by the friction of the blunt steel borer against the bottom of the hollow metallic cylinder, was *greater* than that produced in the combustion of *nine wax candles*, each ¾ inch in diameter, all burning together with clear bright flames.

"One horse would have been equal to the work performed, though two were actually employed. Heat may thus be produced merely by the strength of a horse, and, in case of necessity, this heat might be used in cooking victuals. But no circumstances could be imagined in which this method of procuring heat would be advantageous; for more heat might be obtained by using the fodder, necessary for the support of a horse, as fuel.

* * * * * "It is hardly necessary to add, that anything which any *insulated* body or system of bodies can continue to furnish *without limitation*, cannot possibly be a *material substance;* and it appears to me to be extremely difficult, if not quite impossible, to form any distinct idea of anything capable of being excited and communicated in those experiments, except *motion.*"

Davy.—First scientific memoir, entitled "On Heat, Light, and the Combinations of Light." Works Vol. II.

"*Experiment.*—I procured two parallelopipedons of ice, of the temperature of 29°, six inches long, two wide, and two-thirds of an inch thick; they were fastened by wires, to two bars of iron. By a peculiar mechanism, their surfaces were placed in contact, and kept in a continued and most violent friction for some minutes. They were almost entirely converted into water, which water was collected, and its temperature ascertained to be 35°, after remaining in an atmosphere of a lower temperature for some minutes. The fusion took place only at the plane of contact of the two pieces of ice, and no bodies were in friction but ice.

"From this experiment it is evident that ice by friction is converted into water and according to the supposition, its capacity is diminished; but it is a well-known fact that the capacity of water for heat is much greater than that of ice; and ice must have an absolute quantity of heat added to it before it can be converted into water. Friction consequently does not diminish the capacity of bodies for heat.

* * * * * * *

"Now a motion or vibration of the corpuscules of bodies must be necessarily generated by friction and percussion. Therefore we may reasonably conclude that this motion or vibration is heat, or the repulsive power.

Davy in his *Chemical Philosophy*, p. 95, says:

"* * * * The immediate cause of the phenomena heat, then, is motion, and the laws of its communication are precisely the same as the laws of the communication of motion."

A process similar to the above classical experiment of Davy, has from time immemorial been used by savage nations in obtaining fire by means of friction: not by *rubbing* together sticks, as usually stated, for no one could thus produce ignition, but by *whirling* rapidly, a pointed rod against a wooden block, by means of an arrangement similar to the watchmaker's bow and drill. It is worthy of remark that this apparatus has everywhere the same form, whether used by the islanders of the Pacific or by the aborigines of our own country. One of these instruments can be seen in the State Cabinet of Natural History of New York.

Henry (Dr. Joseph).—Meteorology—Patent Office Report for 1857.

On Acoustics applied to Public Buildings—Smithsonian Report, 1856, p. 228.

"The tuning-fork was next placed upon a cube of India rubber, and this upon the marble slab. The sound emitted by this arrangement was scarcely greater than in the case of the tuning-fork suspended from the cambric thread, and from the analogy of the previous experiments, we might at first thought suppose the time of duration would be great; but this was not the case. The vibrations continued only about forty seconds. The question may here be asked, what became of the impulses lost by the tuning-fork? They were neither transmitted through the India rubber nor given off to the air in form of sound, but were probably expended in producing a change in the matter of the India rubber, or were converted into heat, or both. Though the inquiry did not fall strictly within the line of this series of investigations, yet it was of so interesting a character in a physical point of view to determine whether heat was actually produced, that the following experiment was made.

* * * "And the point of a compound wire, formed of copper and iron, was thrust into the substance of the rubber, while the other ends of the wire were connected with a delicate galvanometer. The needle was suffered to come to rest, the tuning-fork was then vibrated, and its impulses transmitted to the rubber. A very perceptible increase of temperature was the result. The needle moved through an arc of from one to two and a half degrees. The experiment was varied, and many times repeated; the motions of the needle were always in the same direction, namely, in that which was produced when the point of the compound wire was heated by momentary contact with the fingers. The amount of heat generated in this way is, however, small, and indeed, in all cases in which it is generated by mechanical means, the amount evolved appears very small in comparison with the labor expended in producing it."

Leibnitz.—"The *force* of a moving body is proportional to the square of its velocity, or to the height to which it would rise against gravity."

Wollaston.—Bakerian Lecture on the Force of Percussion. Phil. Trans., Vol. XCVI., 1806.

"In short, whether we are considering the sources of extended exertion or of accumulated energy, whether we compare the accumulated forces themselves by their gradual or by their sudden effects, the idea of mechanic force in practice is always the same, and is proportional to the *space* through which any moving force is exerted or overcome, or to the *square* of the velocity of a body in which such force is accumulated."

Tyndall.—Heat considered as a Mode of Motion. New York, 1863.

I need hardly refer the reader to this charming exposition of Prof. Tyndall; who, by his enthusiasm and vividness of illustration, has rendered his subject so popular that his work is, in this country, in the library of nearly every man of culture.

10. *Attraction,* 11. *Repulsion.*

Attraction exists between the minute parts or *atoms* of bodies, and they would continually approach, until contact supervened, if they were not kept apart by an equal and opposing repulsion. These molecular attractions and repulsions are often designated as *the molecular forces.*

The phenomena of traction, of compression, of porosity, and of the transmission of vibrations through all kind of matter, prove indirectly that bodies are formed of minute parts which do not touch, but are kept at certain distances depending on the intensity of the attractions and repulsions subsisting between them; while there are many direct proofs of the above statement which will be given further on.

All *masses* of matter mutually attract each other with an intensity directly proportional to their masses, and inversely as the squares of their distances. This tendency is called *gravitation*, and is common to all matter. This is shown in the celebrated experiment devised by the Rev. John Michell, and generally known as the Cavendish experiment for determining the density of the earth. Describe this apparatus.

Electric and magnetic attractions and repulsions also follow the law of the inverse squares of the distances.

The different manifestations of attraction and repulsion may be thus arranged:

GRAVITATION, ELECTRICITY, MAGNETISM,	Which act at sensible distances; *i. e.* beyond the $\frac{1}{2000}$th of an inch,
COHESION, ADHESION, CAPILLARITY, CHEMICAL AFFINITY,	Which act at insensible distances; *i. e.* within the $\frac{1}{2000}$th of an inch.

Molecular Attraction.

Cohesion designates the attraction existing between the minute parts of the same body; *adhesion* the attraction between the parts of dissimilar bodies. Sometimes designated respectively as homogeneous and heterogeneous attraction.

Experiments on *Cohesion of solids.*—Two leaden planes pressed together, cohere with a force of forty pounds to the square inch. Two plates of glass cohere even in vacuo, which shows that the phenomenon is not due to the atmospheric pressure. The intensity of the cohesion in this experiment is proportional to the surface, and increases with the time of contact. In plate glass manufactories, mirror glasses sometimes cohere with such force, from having been placed on each other without intervening paper, that it is impossible to separate them. It is to be remarked that in the above instances only a comparatively few points of the cohering surfaces are in contact.

"There is a precious experiment by Mr. Huyghens in No. 86 of the Philosophical Transactions. A piece of mirror glass being laid on the table, and another, to which a handle was cemented on one surface, being gently pressed on it, with a little of a sliding motion, the two adhered, and the one lifted the other. Lest this should have been produced by the pressure of the atmosphere, Mr. Huyghens repeated the experiment in an exhausted receiver, with the same success.

"He found that one plate carried the other, although they were not in mathematical contact, but had a very sensible distance between them. He found this by wrapping round one of the plates a single fibre of silk drawn off from the cocoon. The adhesion was vastly weaker than before, but still sufficient for carrying the lower plate.

"Here, then, is a most evident and and incontrovertible example of a mutual attraction acting at a distance. Mr. Huyghens found that if, in wrapping the fibre round the glass, he made it cross a fibre already wrapped round it, there was no sensible attraction. In this case, the glasses were separated by a distance equal to twice the diameter of a fibre of silk.

"I said that this experiment showed that it was not the attraction of gravitation that produced the cohesion. I have repeated the experiment with the most scrupulous care, measuring the disance of the glasses (the diameter of a silk fibre), and the weight supported. I find this, in all cases, to be nearly 14½ times the action of gravity. The calculation is obvious and easy. I tried it in distances considerably different, according to the diameter of the fibre. I must inform the person who would derive his information from his own experiments, that there are many circumstances to be attended to which are not obvious, and which materially affect the result. The silk fibres are not round, but very flat, one diameter being almost double of the other. The 2400th part of an inch may be considered as the average smaller diameter of a fibre. A magnifying glass must be used, and great patience in wrapping the fibre round the glass so that it may not be twisted. A flaxen fibre is much preferable, when gotten single, and fine enough, for it is a perfect cylinder. I must also inform him, that no regularity will be had in experiments with bits of ordinary mirror; these are neither flat enough, nor well enough polished. We must employ the square pieces which are made and finished by a *very few London artists* for the specula of the best Hadley's [Godfrey's] quadrants. These must be most carefully cleaned of all dust or damp. Yet this must not be done by wiping them with a clean cloth; this infallibly deranges everything by rendering the plate electric. I succeeded best by keeping them in a glass jar, in which a piece of moist cloth was lying, but not touching the glasses. When wanted, the glasses are taken out with a pair of tongs and held a little while before the fire, which dissipates the damp which had adhered to them, and which prevented all electricity. With these precautions, and a careful measurement of the diameter of the silk fibre, the experiments will rarely differ among themselves one part in ten.

"* * * * * * If the plates have been hard pressed, with a sliding or grinding motion, the adhesion is then either very strong, or nothing at all; when they do adhere, it seems to be another stage or alternation of the force, as will be explained by and by. But they rarely adhere, owing to fragments torn off by the grinding. The glasses will be scratched by it.

"I thought this capital experiment worthy of a very minute description, it being that which gives us the means of mathematical and dynamical treatment in the greatest perfection." "*A System of Mechanical Philosophy*, by John Robison, L. L. D. Edited by Sir David Brewster, Edinburg: printed for John Murray, London, 1822." Vol. I. page 240, *et seq.*

Cohesion in solids is measured by the force in pounds avoirdupois required to tear apart, by a direct pull, a rod of one square inch area of section. This measure is called the *tenacity* of a body.

Table of Tenacities from "Rankine's Applied Mechanics."

Steel	115,000
Iron, wire	95,000
" wire ropes	90,000
" wrought bars and bolts	65,000
" cast	16,500
Copper, wire	60,000

Copper, cast	19,000
Brass, wire	49,000
" cast	18,000
Gun-metal (copper 8, tin 1)	36,000
Zinc	7,500
Tin, cast	4,600
Lead, sheet	3,300
Teak	18,000
Ash	17,000
Mahogany	16,200
Locust	16,000
Oak, European	14,900
" American red	10.250
Fir—red pine	13,000
" spruce	12,400
" larch	9,500
Chestnut	12,000
Beech	11,500
Maple	10,600
Hempen cables	5,600
Slate	11,200
Glass	9,400
Brick	290
Mortar	50

Cohesion of liquids is shown in the force required to separate a disk of wood from a liquid which wets it; this force varies with the liquid, requiring 52 grains per square inch to separate the disk from water; 31 grains for oil of turpentine; 28 grains for alcohol. These experiments, however, as will be seen below, give only the *relative* cohesion.

Prof. Joseph Henry, in a valuable contribution to molecular physics, published in the *Proceedings of the American Philosophical Society* for April, 1844, showed that the molecular attraction of water for water, instead of being only about fifty-two grains to the square inch, is really several hundred pounds, and is probably equal to that of the attraction of ice for ice. The following are extracts from Dr. Henry's paper:

"The passage of a body from a solid to a liquid state is generally attributed to the neutralization of the attraction of cohesion by the repulsion of the increased quantity of heat; the liquid being supposed to retain a small portion of its original attraction, which is shown by the force necessary to separate a surface of water from water, in the well known experiment of a plate suspended from a scale beam over a vessel of the liquid. It is, however, more in accordance with all the phenomena of cohesion to suppose, instead of the attraction of the liquid being neutralized by the heat, that the effect of this agent is merely to neutralize the polarity of the molecules so as to give them perfect freedom of motion around every imaginable axis. The small amount of cohesion (52 grains to the square inch), exhibited in the foregoing experiment, is due, according to the theory of capillarity of Young and Poisson, to the tension of the exterior film of the surface of water drawn up by the elevation of the plate. This film gives way first, and the strain is thrown on an inner film, which, in turn, is ruptured; and so on until the plate is entirely separated; the whole effect being similar to tearing the water apart atom by atom.

"Reflecting on the subject, the author has thought that a more correct idea of the magnitude of the molecular attraction might be obtained by studying the tenacity of a more viscid liquid than water. For this purpose, he had recourse to soap water, and attempted to measure the tenacity of this liquid by means of weighing the quantity of water which adhered to a bubble of this substance just before it burst, and by determining the thickness of the film from an observation of the color it exhibited in comparison with Newton's scale of thin plates. Although experiments of this kind could only furnish approximate results, yet they showed that the molecular attraction of water for water, instead of being only about 52 grains to the square inch, is really several hundred pounds, and is probably equal to that of the attraction of ice for ice. The effect of dissolving the soap in the water, is not, as might at first appear, to increase the molecular attraction, but to diminish the mobility of the molecules, and thus render the liquid more viscid.

"According to the theory of Young and Poisson, many of the phenomena of liquid cohesion, and all those of capillarity, are due to a contractile force existing at the free surface of the liquid, and which tends in all cases to urge the liquid in the direction of the radius of curvature towards the centre, with a force inversely as this radius. [The explanation of the existence of this contractile force will be given in the next section of the Notes, which considers Capillarity.]

"According to this theory, the spherical form of a dew-drop is not the effect of the attraction of each molecule of the water on every other, as in the action of gravitation in producing the globular form of the planets (since the attraction of cohesion only extends to an appreciable distance), but it is due to the contractile force which tends constantly to enclose the given quantity of water within the smallest surface, namely, that of a sphere. The author finds a contractile force similar to that assumed by this theory, in the surface of the soap-bubble; indeed, the bubble may be considered a drop of water with the internal liquid removed, and its place supplied by air. The spherical form in the two cases is produced by the operations of the same cause. The contractile force in the surface of the bubble is easily shown by blowing a large bubble on the end of a wide tube, say an inch in diameter; as soon as the mouth is removed, the bubble will be seen to diminish rapidly, and at the same time quite a forcible current of air will be blown through the tube against the face. This effect is not due to the ascent of the heated air from the lungs, with which the bubble was inflated, for the same effect is produced by inflating with cold air, and also when the bubble is held perpendicularly above the face, so that the current is downwards.

"Many experiments were made to determine the amount of this force, by blowing a bubble on the larger end of a glass tube in the form of a letter U, and partially filled with water; the contractile force of the bubble, transmitted through the enclosed air, forced down the water in the larger leg of the tube, and caused it to rise in the smaller. The difference of level observed by means of a microscope, gave the force in grains per square inch, derived from the known pressure of a given height of water. The thickness of the film of soap-water which formed the envelope of the bubble, was estimated as before, by the color exhibited just before bursting. The results of these experiments agree with those of weighing the bubble, in giving a great intensity to the molecular attraction of the liquid; equal at least to several hundred pounds to the square inch. Several other methods were employed to measure the tenacity of the film, the general results of which were the same; the numerical details of these are reserved, however, until the experiments can be repeated with a more delicate balance.

"The comparative cohesion of pure water and soap-water was determined by the weight necessary to detach the same plate from each; and in all cases the pure water was found to exhibit nearly double the tenacity of the soap-water. The want of permanency in the bubble of pure water is therefore not due to feeble attraction, but to the perfect mobility of the molecules, which causes the equilibrium, as in the case of the arch, without friction of parts, to be destroyed by the slightest extraneous force."

The above investigation of Dr. Henry will be referred to again under the head of Capillarity.

Gases.—Between the molecules of the same gas repulsion exists, but a slight attraction appears to prevail between the molecules of different gases.

Adhesion.—Of solids to solids.—Placing of metals. Gold leaf stamped on metals. "It is known that if two pieces of metal are *scraped* very clean, a severe blow will make them to cohere so as to be inseparable. It is thus that flowers of gold and silver are fixed on steel and other metals. The steel is first scraped clean, and a thin bit of gold or silver is laid on it, and then the die is applied by a strong blow with a hammer. It is remarkable that they will not adhere with such firmness, if they adhere at all, when the surfaces have been polished in the usual way, with fine powders, &c. This is always done with the help of greasy matters. Some of this probably remains, and prevents that *specific* action that is necessary. I am disposed to think that the scraping of the surfaces also operates in another way, viz: by filling the surface with scratches, that is, ridges and furrows. These allow the air to escape as the pieces come together by the blow. If the mere blow were sufficient, a coin would adhere fast to the die. But, in coining, the flat face of the die first closes with the piece of metal, and effectually confines the air which fills the hollow that is to form the relief of the coin. This air must be compressed to a prodigious degree, and in this state, it is still between the die and the coin. We may say that the impression on the coin is really formed by this included air; for the metal in this part of the coin is never in contact with the die. I know of two cases, which greatly confirm this conjecture. The dies chanced to crack in the highest part of the relief, and after this were thrown aside (although in one, for a common die, the crack was quite insignificant), because the coin could seldom be parted from them."—[*Robison.*]

When bladder is dried on glass, the adhesion is so great that it cannot be torn off without bringing with it some of the glass.

Of solids to liquids.—A rapidly issuing jet of water is deflected from its course by touching a glass rod.—Experiments quoted above, on relative cohesion.

Of liquids to liquids.—Oil and spirits of turpentine spread over the surface of water.

Of gases to solids.—Air and vapor of water adhere with considerable force to the surface of glass. This shown by placing a beaker of water under the receiver of an air pump, when bubbles of air, previously coating as a film the surface of the glass, collect on its surface. That vapor of water also coats with a film the glass, is known from the increase in weight of a dry light glass vessel, when exposed to a damp atmosphere.—The action of clean platinum and gold in condensing gases on their surfaces.—Charcoal absorbs 98 times its volume of ammonia, and 14 volumes of carbonic acid gas. As ammonia is condensed into a liquid by a pressure of seven atmospheres, at a temperature of 60° F., it follows that the absorbed gas must exist in the liquid state in the interstices of the charcoal. Gold leaf will not sink in water from the air condensed on its surface.

Of gases to liquids.—Air and all gas absorbed by water. Oxygen gas absorbed from the air by melted silver.

Of gases to gases.—In the diffusion of gases, one gas acts as a vacuum to another. Vapor diffuses in space containing a gas, until the same tension is produced as would have been acquired by the same vapor evaporating from its liquid in a vacuum.

Molecular Repulsion.

When the convex surface of a plano-convex lens, of a radius of curvature of 20 feet or more, is pressed upon a plate of glass, a system of concentric colored rings are observed. These rings are produced by the interference of certain of the rays of light reflected from the under surface of the lens with those reflected from the upper surface of the glass plate. By knowing the diameter of any ring, and the radius of curvature of the lens, we can calculate the distance between the convex surface and the glass plate corresponding to the ring. Newton thus found that the distance at each ring exceeded the distance of the ring immediately within it by the $\frac{1}{89000}$th of an inch.

Now, unless the lens be heavy, or pressed against the glass plate, no colored spot appears in the centre, and it can be shown that the glasses are, in this case, not in contact, but distant from each other at least $\frac{1}{4450}$th of an inch, and at this distance reposes the upper glass, kept from the plate by a *repulsion* existing between them.

By forcing the glasses nearer together, we at length produce a *black* spot at the centre of the ring system, and Prof. Robison found that "a very considerable force is necessary for producing the black spot. A greater pressure makes it broader, and in all probability this is partly by the mutual yielding of the glasses. I found that before a spot, whose surface is a square inch, can be produced, a force exceeding 1,000 pounds must be employed. When the experiment is made with thin glasses, they are often broken before any black spot is produced.

"What is it that we properly, and without any figure of speech, call a pressure? It is something that we are informed of solely by our sense of touch. What do we feel by means of this sense, when the upper lens lies in our hand? It is not the matter of this lens, for we now see that there is some measurable distance between the lens and the hand; it is this repulsion. Give a blind man a strong magnet in his hand, and let another person approach the north pole of a similar magnet to its north pole. The blind man will think that the other has pushed away the magnet he holds in his hand with something that is soft.

* * * * * * * *

"There is, therefore, an essential difference between *mathematical and physical contact;* between the absolute annihilation of distance and the actual pressure of adjoining bodies. We must grant that two pieces of glass are not in mathematical contact till they are exerting a mutual pressure not less than 1,000 pounds per inch. For we must not conclude that they are in contact till the black spot appears; and even then we dare not positively affirm it. My own decided opinion is, that the glasses not only are not in mathematical contact in the black spot, but the distance between them is vastly greater than the 89,000th part of an inch, the difference of the distances at two successive rings.

* * * * * * * *

"While gravity produces sensible effects at the utmost boundary of the solar system, these [attractions and repulsions] seem limited in their exertion to a small fraction of an inch, perhaps not exceeding $\frac{1}{2000}$th part in any instance; and in this narrow bounds we observe great diversity in the intensity, although we have not yet been able to ascertain the law of variation. What is of peculiar moment, we have seen that those corpuscular forces even change their kind by a change of distance, producing at one distance, the mutual approach, and at another distance the mutual separation of the acting corpuscles, from being attractive, becoming repulsive.

* * * * * * * *

"Physical contact, or *pressure,* becomes *sensible* at the distance of the 5000th part of an inch nearly, and decreases much faster than in the inverse duplicate ratio of the distances. I could infer this from my experiments with the glasses with great confidence, although I could not assign the precise law." *Robison's Mechanical Philosophy*, Vol I. p. 250, *et seq.*

All bodies expand when relieved of pressure, and this expansion is caused by the mutual repulsion of their atoms.

A dew drop does not touch the leaf above which it reposes, but is held at a certain distance by repulsion. Certain insects walk on water, which is repelled by their feet, so that each foot rests in a pit. A needle floats on the surface of water, in which it forms a trough in which it rests.

Prof. Baden Powell has shown (*Phil. Trans.* 1834, p. 485), that the colored rings known as Newton's rings, change their breadth and position in such a manner, when the glasses which produce them are heated, that he inferred that the glasses repelled each other.

"The *distance* at which the repulsive power can act, is shown by these experiments to extend beyond that at which the most extreme visible order of Newton's tints is formed. But I have also repeated the experiment successfully with the colors formed under the base of a prism placed upon a lens of a very small convexity; and according to the analysis of these colors given by Sir John Herschel, the distance is here about the 1100th of an inch."

Very finely divided solids, such as elutriated silica and wood-ashes will, when rendered incandescent, flow like liquids about the capsule which contains them; while it can be directly shown that a sensible distance exists between a layer of these powders and a heated plate on which it rests.

The Molecular Constitution of Matter.—The Atomic Theory of Boscovich.

Of the foregoing facts which we have brought together, concerning the general properties of matter and the effects of attraction and repulsion existing between the minute parts of bodies, we can frame a hypothetical theory which, in a few lines, or postulates, will embrace what we otherwise could not express in many pages.

This theory together with the doctrine of the Conservation of Energy, are the two most important generalizations in Physics, and, in our opinion, the following generalization forms an absolutely essential introduction to the proper study of this department of science.

The ancient Greek philosopher, Democritus, propounded an hypothesis of the constitution of matter, and gave the name of atoms to the ultimate unalterable parts of which he imagined all bodies to be constructed. In the 17th century, Gassendi revived this hypothesis, and attempted to develope it, while Newton used it with marked success in his reasonings on physical phenomena; but the first who formed a body of doctrine which would embrace all known facts in the constitution of matter, was Roger Joseph Boscovich, of Italy, who published at Vienna, in 1759, a most important and ingenious work, styled *Theoria Philosophiæ Naturalis ad unicam legem virium, in Natura existentium redacta.* This is one of the most profound contributions ever made to science; filled with curious and important information, and is well worthy of the attentive perusal of the modern student. In more recent days, the theory of Boscovich has received further confirmation and extension in the researches of Dalton, Joule, Thomson, Faraday, Tyndall and others.

We present here a generalization which, while giving the substance of the important postulates of Boscovich embraces others made necessary by the progress of science since 1760.

1. Matter has trilineal extension.

2. Is impenetrable.

3. Does not form a plenum.

4. All matter consists of indefinitely small but finite parts; of extreme hardness; indivisible, and unalterable by either mechanical or chemical means; and endued with impenetrability and inertia. These ultimate parts are called *Atoms.*

5. These atoms are not in mathematical contact, but are separated from each other by distances which are great when compared with the size of the atoms.

6. A union of atoms forms a molecule, and combinations of molecules form particles of which all bodies are composed.

7. There exist between the atoms, attractions and repulsions: when these tendencies are equal, the atoms preserve fixed positions and the volume of the body is constant. These molecular forces vary both in intensity and direction, by a change of distance, so that at one distance two atoms attract each other, and at another they repel; there being, within the distance in which *physical* con-

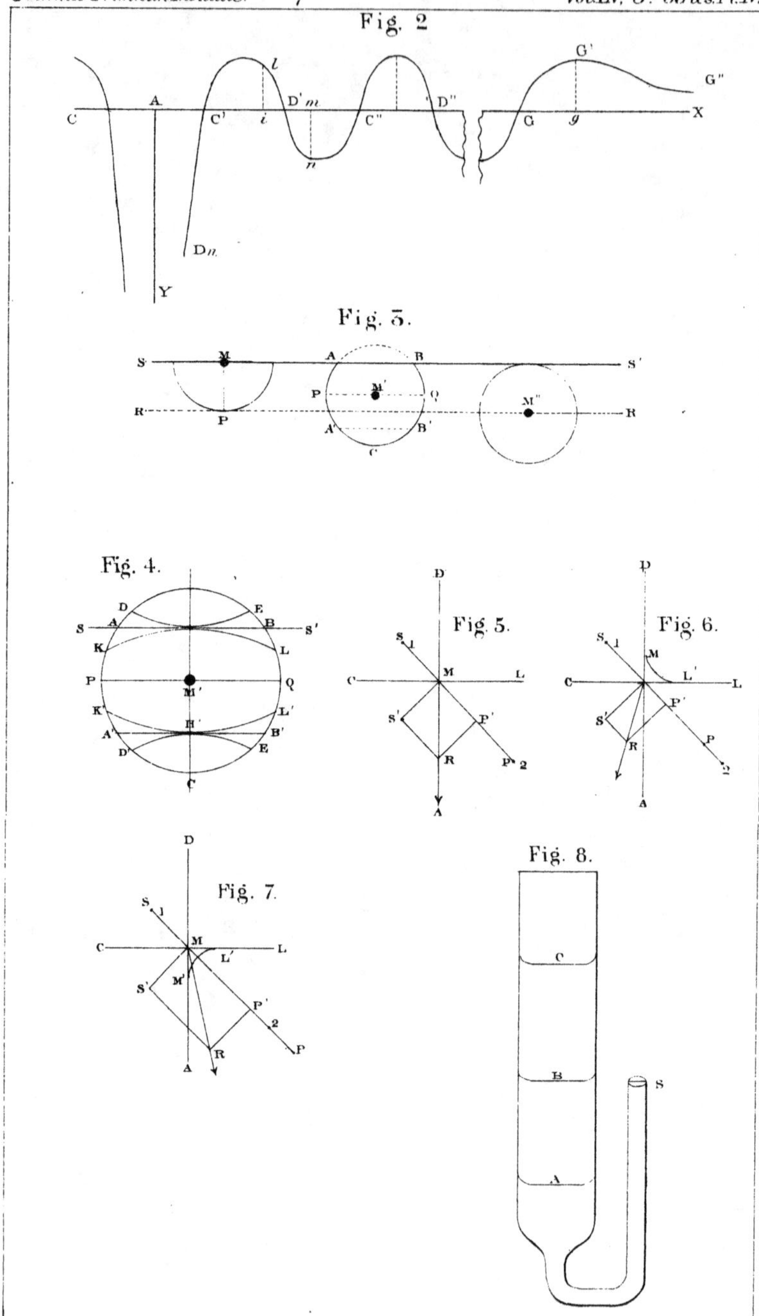
Fig. 2
C
A
C'
i
l
D' m
n
C''
D''
G'
G''
G
g
X
Y
Dn
Fig. 3.
S
M
A
B
S'
P
M'
Q
R
P
A'
B'
C
M''
R
Fig. 4.
D
E
S
A
B
S'
K
L
P
Q
M'
K'
L'
H'
A'
B'
D'
E
C
Fig. 5.
D
S
1
M
C
L
S'
P'
R
P
2
A
Fig. 6.
D
S
1
M
C
L'
L
P'
S'
R
P
2
A
Fig. 7.
D
S
1
C
M
L
L'
M'
S'
P'
2
R
P
A
Fig. 8.
C
B
A
S

tact is observed (about $\frac{1}{2000}$th inch), several alternations of attraction and repulsion.

8. The repulsion of two atoms generally diminishes more rapidly than their attraction when the distance between them is increased; while their repulsion increases more rapidly than their attraction when their distance is diminished.

9. The law of variation is the same in all atoms. It is therefore mutual; for the distance of *a* from *b* being the same as that of *b* from *a*, if *a* attract or repel *b*, *b* must attract or repel *a* with exactly the same force.

10. At all sensible distances (*i. e.* beyond $\frac{1}{2000}$th inch), this mutual tendency is attraction, and varies inversely as the squares of the distances. It is known as *gravitation.*

11. The last force which is exerted between two atoms as their distance diminishes, is an insuperable repulsion, so that no force however great can press two atoms into mathematical contact.

12. Between the molecules of *gases* continued repulsion seems to exist, so that when relieved of exterior force, a gas expands indefinitely.

Between the molecules of *liquids* exist attraction and repulsion, which maintain them at determinate distances; but they have no fixed axial direction, so that a liquid molecule will rotate around any imaginary axis on the action of the slightest force. Thus liquids have *fluidity*, while at the same time they have a small range of compressibility.

In *solids* the molecules have, besides their mutual attractions and repulsions, *polarity* or fixity of position of their axes *inter se*, so that when a molecule in any solid is turned around any axis, it will return to its primitive position after a series of decreasing oscillations.

Those of the above postulates which refer to the mutual action of atoms, can be geometrically expressed by means of an exponential curve. See Plate IV., Fig. 2.

Let an atom be at A, while another is anywhere on the line AX. Suppose that when placed at *i* for example, that an attraction exists between the atoms. The intensity of this attraction is represented by the length of the line *i l*, and we show that mutual attraction exists by drawing the ordinate to the point *i above* the axis AX. If the atom be supposed at *m* and repelled by A, then the repulsion and its intensity are expressed by drawing the ordinate to the point *m*, below the axis AX. This may be supposed to be done for every point of the length AG, which represents the distance between the glass plates in Huyghens' experiment, or about the $\frac{1}{2000}$ inch, and thus we will form an exponential curve.

As there are several alternations of attractions and repulsions, the curve will consist of various inflections lying alternately above and below AX. The last inflection, most distant from A, viz: G′ G″ is of such a form that the lengths of its ordinates being the reciprocals of the squares of their distances from A, it expresses the law of gravitation; the atom at G being at the point called the *limit of gravitation*, or about $\frac{1}{2000}$ inch from A. AX will be an asymptote to this curve, while the inflection C′ D*n* will have AY for asymptote; for the ordinate expressing repulsion increases beyond all limit when the distance from A is just vanishing. The intermediate branches of the curve must be determined by means of the alternations of attraction and repulsion, in the experiments already described and by the aid of the various phenomena of capillarity and of molecular physics.

If an atom, supposed at the point C′, or C″, or &c., have its distance increased from A, it will, being under the curve, be attracted with an intensity represented by its ordinate. When set free it will move with an increasing velocity towards its primitive position of equilibrium, which it will surpass on account of its inertia, and, coming into the sphere of repulsion, it will be repelled from A, and thus oscillate about its point of equilibrium. The atom will therefore eventually return to C′, or C″, or &c. These positions are called *limits of cohesion*, C′ being designated as the *last limit of cohesion.*

If an atom at D′, or D″, or &c., be moved ever so little from its position, it will rush to an adjacent limit of cohesion, either to the right or to the left, according as it was moved from or towards A. These points, D′, D″ &c., are called *limits of dissolution*, and differ from the limits of cohesion in being positions of unstable equilibrium, and therefore only a *temporary molecule* can be formed by an atom placed at D′, D″, &c., with an atom at A; while an atom at C′, C″, &c., together with the atom A forms a *permanent molecule*, which resists compression and dilatation, and whose component atoms return to their primitive positions when the extraneous force is removed; provided the compression or dilatation has not been too great; for, in that case, the atom C″, for example, might be forced beyond D′ by a compression, or removed beyond D″ by a dilatation, and would then rush to another position of permanent equilibrium, either to C′ or to G. The only molecule that cannot possibly be changed by compression is AC′. When, however, the amount of compression or dilatation of a body formed of permanent molecules is a very small fraction of its volume, the body regains the dimension it had before the compression or dilatation was applied, and it is found that the compression or dilatation is proportional to the force employed; for, in this case, the small portion of the curve which expresses the variation of repulsion or attraction may be considered a straight line, and therefore its ordinates are as its abscissas.

These logical consequences of the theory are confirmed by the most extensive experience. "Mr. Coulomb was engaged (for a particular purpose), in a series of experiments on the oscillations of springs, particularly of twisted wires. He suspended a nicely turned ball or cylinder by a wire of a certain length, and fitted it with an index, which pointed out the degrees of the torsion. He found that when a wire of twenty inches long was twisted ten times, the index returned to its primitive position, if repeated a thousand times, and the oscillations were made in equal times, whether wide or narrow. But if it was twisted eleven times, the index did not return to its first place, but wanted nearly a whole turn of it. Here, then, the parts of the wire had taken new relative positions, in which they were again at rest. But what was most remarkable in Coulomb's experiments was this: He found that after the wire had taken this set (as it is termed by the artizans), it exhibited the same elasticity as before. It allowed a torsion of ten turns, and when let go, it returned, and after its oscillations were finished, it rested in the position from which it had been taken. I was much struck with this experiment, and immediately repeated it on a great variety of substances with the same result. The most inelastic substance that I know is soft clay. I got a thread made of fine clay at a pottery, by forcing it through a syringe. It was about $\frac{1}{12}$th of an inch in diameter, and eleven feet long. While quite soft (and smeared with olive oil, to prevent its stiffening by the evaporation of its moisture), I fastened it to the ceiling, and fixed a small weight and an index to its lower end. I found that it made $5\frac{1}{2}$ turns a hundred times and more, without the smallest diminution of its elasticity, always recovering its first position. But when I gave it 7 turns it returned only $5\frac{1}{2}$. Thus it took a set. In this new arrangement of its parts, I found that it again bore a twist of 5½ turns without taking any new set. And I repeated this several times. I then gave it 10 turns, in the same direction with the first seven. It returned 5½ as before, and was again perfectly elastic within this limit."

* * * * * * * *

"Another appearance of tangible matter shows a most encouraging conformity to the theory. Where bodies are very moderately compressed or dilated, the forces employed are proportional to the change of distance between the particles. This appears most exactly true in the experiments of Dr. Hooke, on which he founded his theory of springs, expressed in the phrase *ut tensio sic vis*, and his noble improvement of pocket watches by applying a spiral spring to the axis of the balance, which, by its bending and unbending, produced a force proportional to the angle of the oscillations, and therefore made them isochronous, whether wide or narrow. It is also confirmed by the experiments of Coulomb on twisted wires, and by the form of the elastic curve, as determined by Bemouilli, on the supposition that the forces with which the particles attracted and repelled each other, are proportional to their removal from their natural quiescent positions. But it is found that when the compression or dilatation is too much increased, the resistance does not increase so fast; that it comes to a maximum by still increasing the strain, then decreases, and the body takes a great set or breaks. All this is perfectly analogous to the forces expressed by the ordinates of our exponential curve. In the immediate vicinity of the limits of cohesion, the ordinates increase nearly in the ratio of the abscissæ, then they increase more slowly, come to a maximum, decrease again, till we come to a limit of dissolution."

§ VI. *Capillary Attraction.*

The phenomena of capillary attraction consist in the elevation or depression of the surfaces of liquids along the line of contact with the walls of the vessels which contain them; in the ascent or depression of liquids between slightly separated plates, or in tubes of such small internal diameters as to approach to the dimensions of a hair; whence the name of capillarity, from *capillus*, a hair.

These effects are due to the attractions of the molecules of the liquid for each other combined with the attractions existing between them and the molecules of the solid.

Generalization of the Phenomena of Capillary Attraction.

1. The ascent or depression of the liquid is inversely as the diameter of the tube; provided that this diameter does not exceed two millimêtres. In tubes over twenty millimêtres in diameter, there is neither elevation or depression of liquids.

2. The phenomena are independent of the pressure to which the apparatus is subjected; being the same in vacuo as in compressed air.

3. They do not depend on the thickness of the tube; hence the action of the tube is limited in its effects to insensible distances.

4. The phenomena vary with the material of the tube, and with the nature of the liquid; thus, in a tube of glass, water rises above and mercury is depressed below the level of the outside liquid. The following table of the experiments of M. Frankenheim, gives the heights in millimêtres to which different liquids rise, at a temperature of 0° C., in a glass tube of 1 millimêtre in diameter.

LIQUIDS.	DENSITY.	ELEVATION.
Water....................	1·000	30·73
Formic Acid	1·105	20·40
Acetic Acid.............	1·290	17·02
Sulphuric Acid..........	1·840	16·80
Solution of Potassa...	1·274	15·40
Petroleum	0·847	13·90
Spirits of Turpentine..	0 890	13·52
Acetic Ether............	0·905	12·20
Alcohol	0·821	12·10
Alcohol.................	0·967	14·54
Ether......................	0 737	10·80
Bisulphide of Carbon..	1·290	10·20

5. When the liquid *wets* the tube, it rises above the level of the

liquid outside the tube; and in this case the surface of the elevated liquid is *concave.*

Example. Water in glass tube.

6. When the liquid *does not wet* the tube, it is depressed below the surface of the exterior liquid; and in this case, the surface of the liquid in the tube is convex.

Example. Mercury in tube of glass.

7. When the liquid in the tube has a plane surface, there is neither elevation or depression.

Example. Water in a tube of steel.

These facts are readily explained by the atomic theory, of which they are a beautiful illustration and a natural deduction.

(*a.*) An attraction exists between the neighboring molecules of a liquid, and between the molecules of a liquid and of the contiguous solid.

(*b.*) This force decreases very rapidly as the distance between the molecules increases, and becomes null when that distance exceeds *the radius of sensible attraction.*

(*c.*) The attraction existing between the molecules forming the surface of a liquid, and those extending below the surface as far as the radius of sensible attraction, produces a *molecular pressure*, or tension, on this surface, whose effect has to be added to the pressures produced by gravity and the atmosphere.

(*d.*) The molecular pressure is greater with a convex and less with a concave than with a plane liquid surface.

The truth of the four preceding postulates, is made clear by what follows:

Let S S′, Fig. 3, be a liquid surface of any form. M is a molecule on the surface; M′ is a molecule distant from the surface less than the radius of sensible attraction; and M″ a molecule whose distance from the surface equals the radius of sensible attraction; while all molecules between S S′ and R R′ are distant from the surface less than the radius of sensible attraction.

The molecule, M, on the surface, is attracted downward by all the molecules contained in the portion of sphere which has for its radius M P, the radius of sensible attraction. The effect of all these attractions on M will be a resultant in the direction M P, perpendicular to the surface.

The molecule, M′, is attracted by all the molecules contained in he spherical portion A B C, which we can divide into three parts by three equidistant planes, A B, P Q, A′ B′, parallel to the surface, S S′. The attraction produced by A B P Q, is destroyed by the attraction of P Q A′ B′, and therefore the molecule M′ is drawn downward as though it were attracted only by the liquid contained in A′ B′ C, which gives a resultant, P′, also perpendicular to the surface, but less than P.

The molecule, M″, whose distance from the surface equals the radius of sensible attraction, and all other molecules placed at greater distances, are equally attracted on all sides, and therefore they produce no tension in the surface-film of the liquid, which has for its thickness the radius of sensible attraction.

The influence of the curvature of the liquid surface on the molecular pressure.

Let M′, Fig. 4, be a molecule at a distance M H from the surface S S′ of the liquid. With M as a centre, draw a sphere whose radius M′ P equals the radius of sensible attraction.

If the surface is a plane, A B, the attractions of the liquid in A B P Q are destroyed by those produced by the symmetrical portion below, A′ B′ P Q, and there remains for resultant only the action of A′ B′ C.

Suppose the surface concave and D H E; if we draw through H′ the symmetrical surface, D′ H′ E′, it is evident that the attractions of the molecules comprised between D H E P Q and of those contained within D′ H′ E′ P Q equal and oppose each other, and there remains only the attraction of D′ H′ E′ C on M′, which is less than when the surface was a plane.

If the surface is convex, and is represented by K H L, draw the symmetrical surface, K′ H′ L′; then the efficient attracting portion of the liquid will be increased and represented by K′ G′ L′, and consequently the molecular pressure is greater with a convex than with a plane surface.

We can now explain the rise and depression of liquids in capillary tubes.

When the surface of the liquid in the tube is *concave*, the molecular pressure on the liquid in the tube is less than the pressure on the liquid outside the tube, and therefore the liquid rises in the tube to a height which measures the diminution of pressure produced by the concave surface.

When the surface of the liquid in the tube is *plane*, there is neither elevation or depression, for the pressures are the same on the surfaces of the liquid inside and outside the tube.

When the surface of the liquid in the tube is convex, the molecular pressure on the liquid in the tube is more than the pressure on the liquid outside the tube, and therefore the liquid column is depressed in the tube below the level of the outside liquid, and the depth to which the column is forced below this level, is the measure of the pressure produced by the convex surface.

As we have seen that the elevation or depression of liquids in capillary tubes, is due to a diminution or increase of molecular pressure, produced by a concave or convex surface, it remains, to render the explanation complete, to show the cause of the special figure of each surface.

Cause of the (1) plane, (2) concave, and (3) convex surfaces of liquids in capillary tubes.

Let D A in Figs. 5, 6 and 7, be the vertical surface of a solid plunged in liquids, whose surfaces are M L. Let M be a molecule of the surface of the liquid contiguous to the plate. This molecule is attracted by all the molecules contained in the quarter-spheres D M C and A M C, whose radii are equal to the distance of sensible attraction; giving as resultants M S and M S′, while the resultant of the attractions of the liquid on the molecule, M, will be M. P.

Three cases can present themselves.

1. If the resultant, M P, Fig. 5, is *twice* M S, or its equal, M S′, the effect of these three attractions on M will be the resultant, M R; which being perpendicular to the liquid surface, the fluid will remain *horizontal*, for the surface of a liquid is always perpendicular to the forces acting on it.

2. If the resultant, M. P, Fig. 6, is *less than twice* M S or M S′, the three attractions will result in M R, which will produce a *concave* surface M L′, inclined against the plate.

3. If the resultant, M P, Fig. 7, is *more than twice* M S or M S′, the resultant of the three attractions on liquid contiguous to solid will be M R, which will, for the reason given above, produce the surface M′ L′. which will be *convex*.

The above results may be expressed concisely as follows:

I. On the free surface of every liquid there exists a molecular pressure from without inward, which always adds its effect to that produced by gravity and the pressure of the air.

II. The intensity of this molecular pressure varies with the form of the surface, being greater when the surface is convex and less when concave, than when it is plane.

III. The form of a liquid surface in a tube, depends on the relative amounts of attraction existing between the molecules of the liquid and the molecules of the solid and of the liquid.

1. When the attraction between the molecules of the liquid is *twice* as great as the attraction between the molecules of the liquid and those contained in an equal portion of the solid, the surface in the capillary tube is *horizontal.*

2. When the attraction between the molecules of the liquid is *less than twice* that existing between the molecules of the liquid and solid, the surface in the tube is *concave.*

3. When the attraction between the molecules of the liquid *is more than twice* that between the molecules of the liquid and solid the surface in the tube is *convex.*

IV. When the surface of the liquid in the capillary tube is (*a*) *horizontal*, it is in the same plane with the exterior liquid. (*b*) *concave*, it is above the plane of the exterior liquid. (*c*) *convex*, it is below the plane of the exterior liquid.

V. The amount of elevation or of depression of the same liquid in tubes of the same material, is inversely as the diameter of these tubes. This is known as the law of Jurin, after the philosopher who established it; and with the aid of the table already given we can by means of it readily calculate the heights to which different liquids will rise in glass tubes of various dimensions, contained within diameters of two millimêtres to a few hundredths of a millimêtre.

The reason of this law is as follows. The force which elevates or depresses the liquid columns in the tubes depends evidently, from what has preceded, upon the number of the molecules on the surface of the liquid contiguous to the sides of the tubes. Therefore, the forces of elevation or of depression are as the interior circumferences of the tubes, and the forces are measured by the quantity (or weight) of liquid elevated above or depressed below the level of the liquid exterior to the capillary tube. Therefore, let h and h' be the lengths of liquid columns elevated or depressed in tubes whose interior diameters are respectively d and d'. Their interior circumferences are πd and $\pi d'$. δ being the specific gravity of the liquid, the weights of the columns elevated or depressed will be $\frac{1}{4} \pi d^2 h \delta$, and $\frac{1}{4} \pi d'^2 h' \delta$. These weights are equal to the forces which produce the elevations or depressions of the liquid columns, and these forces being to each other as the interior circumferences of the tubes, we have

$$\pi d : \pi d' :: \tfrac{1}{4} \pi d^2 h \delta : \tfrac{1}{4} \pi d'^2 h' \delta$$

or

$$d : d' :: h' : h$$

which is the expression of the law given above.

Experiments.—The apparatus with which Gay Lussac verified the above law, explained and used.

If two squares of plane glass, touching along two vertical edges, are opened to an acute angle and placed in colored water, the liquid will rise between the plates, forming an equilateral hyperbola, and therefore the liquid at various points stands at heights inversely as the distance of the plates at these points.

The relation which exists between the form of the surface which terminates the capillary column and its vertical distance above the plane of the exterior liquid, is beautifully shown by the following experiment, which, with those above cited, can be readily thrown on a screen by means of the lantern and erecting prism of Prof. Morton (see *Journal of Franklin Institute*, Vol. LIII., p. 406). A large glass tube has connected with it a capillary tube, as shown in Fig. 8. Water, colored with carmine, is poured into the larger tube until its level reaches, say, A, and the liquid in the capillary tube just attains the top, S, and in these circumstances, will there form a *concave* surface. Now, on pouring into the large tube more liquid, the concave surface becomes flatter and flatter as the liquid rises in the tube A, until, when the surface rises to B on the same level as S, the terminal surface at S is *a plane*. When liquid is further added until the surface reaches C, at a higher level than S, the capillary surface at S is *convex*.

When S is concave, the molecular pressure on this surface is less than on A by the pressure of the column from the level A to S. When S is plane, equality of pressure exists in both tubes, and therefore the liquid surfaces are in the same plane. When S is convex, more molecular pressure is on S than on C, by the column from S to the level C,

Professor Plateau, of the University of Ghent, has made a series of very important investigations in molecular physics, which are contained in a series of papers entitled, "*Experimental and Theoretical Researches on the Figures of Equilibrium of a Liquid Mass withdrawn from the action of Gravity*," translated and published by the Smithsonian Institution, in the Reports of 1863, *et seq.* The fifth series of these investigations (Smith. Rep. 1865), contains a research on the molecular pressure exerted by liquid films, with applications to capillary action; and so interesting has this investigation appeared to us, that we thought it proper to present a rather full abstract from Prof. Plateau's paper.

Pressure exerted by a spherical film on the air which it contains.—Application.

"The exterior surface of a laminar sphere being convex in every direction, the pressure which corresponds to it is greater than that of a plane surface, and consequently the resultant of the pressures exerted in any point of the bubble by the two surfaces of the latter, is directed towards the interior; whence it results that the bubble presses on the air which it encloses. It is, indeed, well known that when a soap-bubble has been inflated, and while it is still attached to the tube, if the other extremity of this last be left open, the bubble gradually collapses, expelling the air which it contained through the tube. We see now what is the precise cause of this expulsion.

"But we may go further, and determine according to what law it is, that the pressure, exerted by such a bubble on the confined air, depends on the diameter of that bubble. We can compute, moreover, the exact value of the pressure in question for a bubble having a given diameter, and formed of a given liquid. The pressure corresponding to a point of a laminar figure, has for its expression $A\left(\frac{1}{R}+\frac{1}{R'}\right)$. R and R′, standing for the radii of curvature, P being the pressure which a plane surface would occasion, and A a constant which depends on the nature of the liquid. Now, in the case of the spherical figure, we have $R = R' =$ the radius of the sphere. If, therefore, we designate by d the diameter of the bubble, the value of the pressure will simply become $\frac{4A}{d}$, always, be it understood, neglecting the slight thickness of the film; whence it follows that the intensity of the pressure exerted by a laminar spherical bubble on the air which it confines, is in inverse ratio to the diameter of that bubble.

"This first result established, let us recur to the general expression of the pressure corresponding to any point of a liquid surface, an expression which is $P + \frac{A}{2}\left(\frac{1}{R}+\frac{1}{R'}\right)$. For a surface of convex spherical curvature, if we designate by d the diameter of the sphere to which this surface pertains, the above expression becomes $P + \frac{2A}{d}$, and for a spherical surface of concave curvature pertaining to a sphere of the same diameter, we shall have $P - \frac{2A}{d}$. Thus, in the case of the convex surface the total pressure is the sum of two forces acting in the same direction—force, of which one designated by P is the pressure which a plane surface would exert, and the other represented by $\frac{2A}{d}$ is the action which depends on the curvature. On the contrary, in the case of the concave surface the total pressure is the difference between two forces acting in opposite directions, and which are again, one the action P of a plane surface, and the other $\frac{2A}{d}$, which depends on the curvature. Whence it is seen that the quantity $\frac{4A}{d}$, which represents the pressure exerted by a spherical film on the air it encloses, is equal to double the action which proceeds from the curvature of one or the other surface of the film.

"Now, when a liquid rises in a capillary tube, and the diameter of this is sufficiently small, we know that the surface which terminates the column raised does not differ sensibly from a concave hemisphere, whose diameter is consequently equal to that of the tube. Let us recall, moreover, a part of the reasoning by which we arrive, in the theory of capillary action, at the law which connects the height of the column raised with the diameter of the tube. Let us suppose a pipe, excessively slender, proceeding from the lowest point of the hemispheric surface in question, descending vertically to the lower orifice of the tube, then bending horizontally, and finally rising again so as to terminate vertically at a point of the plane surface of the liquid exterior to the tube. The pressures corresponding to the two orifices of this little pipe will be, on the one part, P, and on the other, $P - \frac{2A}{\delta}$, if by δ be designated the diameter of the concave hemisphere, or, what amounts to the same thing, that of the tube. Now, the two forces P, mutually destroying one another, there remains only the force $-\frac{2\,A}{\delta}$, which, having a sign contrary to that of P, acts consequently from below upwards at the lower point of the concave hemisphere, and it is this which sustains the weight of the molecular thread contained in the first branch of the little pipe between the point just mentioned and a point situated at the height of the exterior level. This premised, let us remark that the quantity $\frac{2\,A}{\delta}$ is the action which results from the curvature of the concave surface. The double of this quantity or $\frac{4\,A}{\delta}$, will therefore express the pressure exerted on the enclosed air by a laminar sphere or hollow bubble of the diameter δ, and formed of the same liquid. It thence results that this pressure constitutes a force capable of sustaining the liquid at a height double that to which it rises in the capillary tube, and that, consequently, it would form an equilibrium to the pressure of a column of the same liquid having that double height. Let us suppose, for the sake of precision, δ equal to a millimètre, and designate by h the height at which the liquid stops in a tube of that diameter. We shall have this new result, that the pressure exerted on the enclosed air by a hollow bubble formed of a given liquid and having a diameter of 1 millimètre, would form an equilibrium to that exerted by a column of this liquid of a height equal to $2\,h$. Now, the pressure exerted by a bubble being in inverse ratio to the diameter thereof, it follows that the liquid column which would form an equilibrium to the pressure exerted by a bubble of any diameter whatever, d, will have a height equal to $\frac{2h}{d}$.

"It would seem, at first, that this last expression ought to apply equally well to liquids which sink in capillary tubes, h then designating this subsidence, the tube still being supposed 1 millimètre in diameter; but is not altogether so, for that would require, as is readily seen by the reasonings which precede, that the surface which terminates the depressed column in the capillary tube should be sensibly a convex hemisphere; now we know that in the case of mercury this surface is less curved; according to the observations of M. Bède, its height is but about half of the radius of the tube; whence it follows that the valuation of the pressure yielded by our formula would be too small in regard to such liquids. It may be considered, however, as a first approximation.

Let us take, as a measure of the pressure exerted by a bubble, the height of the column of water to which it would form an equilibrium. Then, if δ designates the density of the liquid of which the bubble is formed, that of water being 1, the heights of the columns of water and of the liquid in question which would form an equilibrium to the same pressure will be to one another, in the inverse ratio of the densities, and, therefore, if the height of the second is $\frac{2h}{d}$, that of the first will be $\frac{2h\delta}{d}$. Hence, designating by p the pressure exerted by a laminar sphere on the air which it encloses, we obtain definitely $p=\frac{2h\delta}{d}$, δ being, as we have seen, the density of the liquid which constitutes the film, h the height to which this liquid rises in a capillary tube 1 millimètre in diameter, and d the diameter of the bubble. If, for example, the bubble be formed of pure water, we have $\delta=1$, and, according to the measurements taken by physicists, we have, very exactly, $h=30^{mm}$; the above formula, therefore, will give, in this case, $p=\frac{60}{d}$. If we could form a bubble of pure water of one decimetre or 100^{mm}, in diameter, the pressure which it would exert would consequently be equal to $0^{mm}\cdot6$, or, in other terms, would form an equilibrium to the pressure of a column of water $0^{mm}\cdot6$ in height; the pressure exerted by a bubble of the same liquid one centimetre, or 10^{mm} in diameter, would form an equilibrium to that of a column of water 6^{mm}. As regards soap-bubbles, their pressures, if the solution were as weak as possible, would differ very little from those exerted by bubbles of the same diameters formed of pure water.

"For mercury we have $\delta=13\cdot59$, and, according to M. Bède, h about equal to 10^{mm}; the formula would therefore give, for a bubble of mercury $p=\frac{271\cdot8}{d}$, but, from the remark which closes the last paragraph, this value is too weak, and can only be regarded as a first approximation. It only instructs us that, with an equality of diameter, the pressure of a bubble of mercury would exceed four and a half times that of a bubble of pure water. For sulphuric ether, we have $\delta=0\cdot715$, and conclude from measurements taken from M. Frankenheim, h to be very closely to $10^{mm}\cdot2$; whence results $p = \frac{14{\cdot}6}{d}$, and thus, with an equal diameter, the pressure of a bubble of sulphuric ether would be but the fourth of that of a bubble of pure water.

We know that the product $h\delta$ being the product of the capillary height by the density, is proportional to the molecular attraction of the liquid for itself, or, in other terms, to the cohesion of the liquid; (see research of Prof. Henry on Cohesion of Liquids, quoted in §V.) it is, moreover, the result from a comparison of the values $\frac{4\,A}{d}$ and $\frac{2\,h\delta}{d}$, which have been found to represent the pressure exerted by a laminar sphere on the air which it contains; hence we deduce $h\delta = 2\,A$, and it will be remembered that A is the capillary constant; that is to say, a quantity proportional to the cohesion of the liquid. The formula $p = \frac{2\,h\delta}{d}$ indicates, therefore, as must be evident, that the pressure exerted by a laminary bubble on the included air is in the direct ratio of the cohesion of the liquid which constitutes the film and the inverse ratio of the diameter of the bubble.

"As early as 1830, a learned American, Dr. Hough, had sought to arrive at the measure of pressure exerted, whether on a bubble of air contained in an indefinite liquid or on the air enclosed in a bubble of soap.

(*Inquiries into the principles of liquid attraction.*—Silliman's Journal, 1st series, vol. xvii., page 86.)

"He conceives quite a just idea of the cause of these pressures which he does not, however, distinguish from one another, and, in order to appreciate them, sets out, as I have done, with a consideration of the concave surface which terminates a column of the same liquid raised in a capillary tube; but, although an ingenious observer, he was deficient in a knowledge of the theory of capillary action, and hence arrives, by reasoning, of which the error is palpable, at values and a law which are necessarily false.

"Prof. Henry, in a very remarkable verbal communication on the cohesion of liquids, made in 1844, to the American Philosophical Society, (Philosophical Magazine, 1845, vol. xxvi., page 541), described experiments by means of which he had sought to measure the pressure exerted on the internal air by a bubble of soap of a given diameter. According to the account rendered of this communication, the mode of operation adopted by Mr. Henry was essentially as follows: he availed himself of a glass tube of **U** form, of small interior diameter, one of whose branches was bell-shaped at its extremity, and inflated a soap-bubble extending to the edge of this widened portion; he then introduced into the tube a certain quantity of water, and the difference of level in the two branches gave him the measure of the pressure. Unfortunately, the statement given does not make known the numbers obtained, nor does it appear that Mr. Henry has subsequently published them. This physicist refers the phenomenon to its real cause, and states the law which connects the pressure with the diameter of the bubble; the account does not say whether the experiments verified it. But Mr. Henry considers that a hollow bubble may be assimilated to a full sphere reduced to its compressing surface; that is to say, he attributes the phenomenon to the action of the exterior surface of the bubble, without taking into account that of the interior surface. Let us add that, in the same communication, Mr. Henry has mentioned several experiments which he had made on the films of soap and water, and which, from the statement given would elucidate, in a remarkable manner, the principles of the theory of capillary action. It is much to be regretted that these experiments are not described

"In a memoir presented to the Philomatic Society in 1856, and printed in 1859 in the *Comptes Rendus*, (tome xlviii., page 1405.) M. de Tessan maintains that if the vapor which forms clouds and fogs were composed of vesicles, the air enclosed in a vesicle of 0·02 millimètre diameter would be subjected, on the part of this vesicle, to a pressure equivalent to $\frac{1}{7}$ of an atmosphere. M. de Tessan does not say in what manner he obtained this valuation; but it is easily seen that he has fallen into an error analogous to that of Professor Henry, in the sense that he pays no attention except to the exterior surface of the liquid pellicle. According to the formula of the preceding paragraph, the pressure exerted on the interior air by a bubble of water of 0·02 millimètre diameter would, in fact, be equivalent to that of a column of water 3 metres in height, which equals nearly $\frac{2}{7}$ of the atmospheric pressure; M. de Tessan has found then but half the real value, and we know that this half is the action due to the curvature of one only of the surfaces of the film.

"After having obtained the general expression of the pressure exerted by a laminar sphere on the air which it encloses, it remained for me to submit my formula to the control of experiment. I have employed, with that view, the process of Mr. Henry, which means that the pressure was directly measured by the height of the column of water to which it formed an equilibrium.

From our formula we deduce $p\,d = 2\,h\,\vartheta$; for the same liquid, and at the same temperature, the product of the pressure by the diameter of the bubble must, therefore, be constant, since h and ϑ are so. It is this constancy which I have first sought to verify for bubbles of glyceric liquid—(a solution of Marseilles soap and glycerine in distilled water, with which Plateau made his bubbles)—of different diameters."

Plateau here describes his apparatus and the precautions to be used in making the measures, which the reader will find detailed in the Smithsonian Report for 1865.

"The following table contains the results of these experiments; I have arranged them, not in the order in which they were obtained, but in the ascending order of the diameters, and I have distributed them into groups of analogous diameters. During the continuance of the operations the temperature varied from 18°·5 to 20° C.

* * * * As the first diameter is to those of the last group very nearly as 1 to 6, these results suffice, I think, to establish distinctly the constancy of the product $p\,d$, and consequently the law according to which the pressure is in the inverse ratio of the diameter.

* * * * * * * *

"As to the general mean 22·75 of the results of the table, its decimal part is necessarily a little too high, on account of the excessive

Diameters, or values of d.	Pressures, or values of p.	Products, or values of $p\,d$.
m m	m m	
7·55	3·00	22·65
10·37	2·17	22·50
10·55	2·13	22·47
23·35	0·98	22·88
26·44	0·83	21·94
27·58	0·83	22·89
46·60	0·48	22·37
47·47	0·48	22·78
47·85	0·43	20·57
48·10	0·55	26·45

value 26·45 of the last product. As this product, and that which precedes it, are those which alone deviate materially from 22 in this integral part, it will be admitted, I think, that a nearer approach to the true value will be made by neglecting these two products and taking the mean of the others, a mean which is 22·56, or more simply 22·6; we shall adopt, then, this last number for the value of the product $p\,d$ in regard to the glyceric liquid.

It remained to be verified whether this value satisfied our formula, according to which we have $p\,d = 2\,h\,\vartheta$, the quantities ϑ and h being respectively, as we have seen, the density of the liquid and the height which this liquid would attain in a capillary tube 1 millimètre in diameter. With this view, therefore, it was necessary to seek the values of these two quantities in reference to the glyceric liquid. The density was determined by means of the aerometer of Fahrenheit, at the temperature of 17° C., a temperature little inferior to that of the preceding experiments, and the result was $\vartheta = 1{\cdot}1065$. To determine the capillary height the process of Gay Lussac was employed, that is, the measurement by the cathetometer, all known precautions being taken to secure an exact result. The experiment was made at the temperature of 19° C. * * * * * * * The reading of the cathetometer gave, for the distance from the lowest point of the concave meniscus to the exterior level, $27^{mm}{\cdot}35$.

This measurement having been taken, the tube was removed, cut at the point reached by the capillary column, and its interior diameter at that point measured by means of a microscope, furnished with a micrometer, giving directly hundredths of a millimètre. It was found that the interior section of the tube was slightly elliptical, the greater diameter being $0^{mm}{\cdot}374$ and the smaller $0^{mm}{\cdot}357$; the mean was adopted, namely, $0^{mm}{\cdot}3655$, to represent the interior diameter of the tube assumed to be cylindrical. To have the true height of the capillary column, it is necessary, we know, to add to the height of the lowest point of the meniscus the sixth part of the diameter of the tube, or, in the present case, $0^{mm}{\cdot}06$; the true height of our column is consequently $27^{mm}{\cdot}41$. Now, to obtain the height h to which the same liquid would rise in a tube having an interior diameter of exactly a millimètre, it is sufficient, in virtue of the known law, to multiply the above height by the diameter of the tube, and thus we find definitively $h = 10^{mm}{\cdot}018$.

"I should here say for what reason I have chosen for the experiment a tube whose interior diameter is considerably less than a millimètre. The reasoning by which I arrived at the formula supposes that the surface which terminates the capillary column is hemispherical; now that is not strictly true, but in a tube so narrow as that which I have employed, the difference is wholly inapplicable, so that in afterwards calculating, by the law of the inverse ratio of the elevation to the diameter, the height for a tube one millimètre in diameter, we would have this height such as it would be if the upper surface were exactly hemispherical.

"The values of ϑ and h being thus determined, we deduce therefrom $2\,h\,\vartheta = 22{\cdot}17$, a number which differs but little from 22·56 obtained above as the value of the product $p\,d$. The formula $p\,d = 2\,h\,\vartheta$ may therefore be regarded as verified by experiment, and the verification will appear still more complete if we consider that the two results are respectively deduced from elements altogether different. I hope hereafter to obtain new verifications with other liquids.

Investigation of a very small limit below which is found, in the glyceric liquid, the value of the radius of sensible activity of the molecular attraction.

"The exactness of the formula $p = \frac{2\,h\,\vartheta}{d}$ supposes, as we are about to show, that the film which constitutes the bubble has, at all points, no thickness less than double the radius of sensible activity of the molecular attraction.

"We have seen that the pressure exerted by a bubble on the air which it encloses is the sum of the actions separately due to the curvatures of its two faces. On the other hand, we know that, in the case of a full liquid mass, the capillary pressure exerted by the liquid on itself emanates from all points of a superficial stratum having as its thickness the radius of activity in question. Now, if the thickness of the film which constitutes a bubble is everywhere superior or equal to double that radius, each of the two faces of the film will have its superficial stratum unimpaired, and the pressure exerted on the enclosed air will have the value indicated by our formula. But if, at all its points, the film has a thickness inferior to or double this radius, the two superficial strata have not their complete thickness, and the number of molecules comprised in each of them being thus lessened, these two strata must necessarily exert actions less strong, and consequently the sum of these, that is to say, the pressure on the interior air, must be smaller than the formula indicates it to be. Hence it follows that if, in the experiments described above, the thickness of the films which formed the bubbles had, through the whole extent of these last, descended below the limit in question, the results would have been too small, but in this case we should have remarked progressive and continued diminutions in the pressures, which, however, never happened, although the color of the bubbles evinced great tenuity. But all physicists admit that the radius of sensible activity of the molecular attraction is excessively minute.

"But what precedes permits of our going further, and deducing from experiment a datum on the value of the radius of sensible activity, at least in the glyceric liquid. * * *
* * * * * * * *

"After the film has acquired a uniform thinness, if the pressure exerted on the air within the bubble underwent a diminution, this would be evinced by the manometer, and it would be seen to progress in a continuous manner in proportion to the ulterior attenuation of the film. In this case the thickness which the film had when the diminution of pressure commenced would be determined by the tinge which the central space presented at that moment, and the half of that thickness would be the value of the radius of sensible activity of the molecular attraction. If, on the contrary, the pressure remains constant until the disappearance of the bubble, we may infer from the tint of the central space the final thickness of the film, and the half of this thickness will constitute at least a limit, very little below which is to be found the radius in question.
* * * * * * * *

* * * I deposited in the bottom of the dry jar, morsels of caustic potash, and contrived by the application of a little lard around the orifice of the jar and of the aperture through which passed the copper tube, that after the introduction of the bubble, the pasteboard disk should close the opening hermetically. * *
* * * * * * * *

"Now, under these conditions, the diminution of thickness of the film was continuous, the bubble lasted for nearly three days, and when it burst, it had arrived at the transition from the yellow to the white of the first order; it then presented a central space of a pale yellow tint, surrounded by a white ring.

* · * * * * * * *

* * * if the pressure varied, it was in an irregular manner, in both directions, and terminating not in a diminution, but an augmentation, at least relative; we may, therefore, admit, I think, that the final thickness of the film was still superior to double the radius of sensible activity of the molecular attraction.

"Let us now see what we may deduce from this last experiment. According to the table given by Newton, the thickness of a film of pure water which reflects the yellow of the first order is, in millionths of an English inch, $5\frac{1}{3}$, or 5·333, and for the white of the same order $3\frac{7}{8}$, or 3·875. We may therefore take the mean, namely 4·064, as the closely approximative value of the thickness corresponding, at least in the case of pure water, to the transition between those colors, and the English inch being equal to 25·4 millimètres, this thickness is equivalent to $\frac{1}{8554}$ of a millimètre. Now we know that, for two different substances, the thickness of the films which reflect the same tint is in the inverse ratio of the indices of refraction of those substances. In order, therefore, to obtain the real thickness of our film of glyceric liquid, it suffices to multiply the denominator of the preceding fraction by the ratio of the index of the glyceric liquid to that of water. I have measured the former approximatively by means of a hollow prism, and have found it equal to 1·377. That of water being 1·336, there results, for the thickness of the glyceric film $\frac{1}{8811}$ of a millimètre. The half of this quantity, or $\frac{1}{17622}$ of a millimètre, constitutes, therefore, the limit furnished by the experiment in question. Hence we arrive at the very probable conclusion, that in the glyceric liquid the radius of sensible activity of the molecular attraction is less than $\frac{1}{17000}$ of a millimètre.

"I had proposed to continue this investigation with a view to reach, if possible, the black tint, and to elucidate the variations of the manometer; but the cold season has intervened, diminishing the persistence of the bubbles, and I have been forced to postpone attempts to a more favorable period."

In the report of the transactions of the Society of Physics and Natural History, of Geneva, 1862, we find the following: "Prof. Wartmann, Jr., repeated before the Society the recent experiments of M. Plateau on bubbles of soap, of varied forms as well as much persistency, obtained by mixing with soap-suds a small quantity of glycerine, and causing the bubbles to attach themselves to iron wires arranged in different manners. At a subsequent session M. Wartmann exhibited an apparatus of the same kind, still more varied, so as to produce more perfectly than by former processes the phenomena of coloration in extremely thin surfaces of the liquid. The dark part presents not more than $\frac{1}{104920}$ of a millimètre, whence we may conclude, says M. Wartmann, that the radius of the sensible activity of molecular attraction is below $\frac{1}{209000}$ of a millimètre."

ADDITIONS AND CORRECTIONS.

Page 3, line 11 from top, *for* such are the planets, &c., *read* such are the stars, the sun, the planets and asteroids of, &c.

Page 3, after line 21, *read*: Nebulæ are faintly luminous aggregations of matter so far removed from our solar system as to have no sensible parallax. They are formed either of clusters of stars, and are then *resolvable*, or consist of self-luminous gaseous matter, and are then unresolvable into star points; while some seem formed of stars surrounded by atmospheres of intensely heated vapors, as is the case with our sun. These facts are the results of the spectroscopic analysis of these bodies. If the spectrum of a celestial body is formed only of *isolated bright lines*, then that body is composed only of luminous gas, as is the case with the great nebula in Orion. When the spectrum is *continuous*, the light comes from an incandescent solid or liquid mass; while if the spectrum is *continuous and crossed by fine dark lines*, we know that the light emanates from an incandescent solid or liquid body surrounded by vapors which have, by their absorption, produced these dark lines.

Of 60 nebulæ examined by Mr. Huggins, the spectroscope showed that 20 were gaseous, while 40 consisted of aggregations of stars.

(See *Results of Spectrum Analysis Applied to the Heavenly Bodies*. By Wm. Huggins. London: 1868.)

The Zodiacal light is a faint nebulous light, resembling that of the tail of a comet. It is visible in the W. about 1st March, after sunset, and in the

E. about the middle of October, before sunrise. It is supposed to be caused by a nebulous envelope which surrounds the sun and extends beyond the orbit of the earth, so that, if the sun were seen from one of the stars exterior to our system, it would appear as a bright point enveloped in a haze, just as some of the fixed stars appear to us.

Page 4, line 3 from top, *for* solids, liquids and gases, *read* solid, liquid and gaseous.

Page 4, line 8 from bottom, *for* stone, associated facts. (1) Spaces *read* stone. Associated facts, (1) spaces.

Page 17, line 8 from top, *for* $\frac{1}{1000}$ *read* $\frac{1}{4000}$.

Page 18, line 7 from top, *after* audience, *read*, by the aid of a lantern provided with a small slit and a lens.

Page 19, *after* line 3 from top, *read*: See papers on Nobert's Test-plates, by Mr. Stodder, in the *American Naturalist* for April, 1868, and by Messrs. Sullivant and Woodward in *Amer. Jour. Science*, Nov., 1868.

Page 20, *after* line 17 from top, *read:* An arc of 17° 10′, of which the cord is $\frac{1}{8}$ of radius, may be employed as a test of the accuracy of the work.

Page 24, line 2 from top, *for* millimêtre *read* milligramme.

Page 24, line 16 from bottom, *for* mean revolution, *read* mean apparent revolution.

Page 25, line 1 from top, *for dead-beat*, *read dead beat* and *gravity*.

Page 25, *after* line 4 from bottom, *read* Dr. Locke, of Cincinnati, in 1848, first recorded transits by means of an electric clock and the ordinary Morse telegraph register; and Messrs. Saxton and Bond subsequently replaced the telegraph fillet by a uniformly revolving cylinder. This mode of registering transits is known as "the American method."

Page 28, add to the table Litre = 0·035317 cubic feet.
" = 61·02705 " inches.

Page 31 and 32. In this discussion we have assumed that the arithmetical mean is the mean of an infinite number of observations, and, therefore, that its error is an infinitesimal; but, as the number of observations is always limited, better approximations to the values of the probable errors of a series of directly observed quantities are given by the formulæ:—

$$e = \sqrt{\frac{\varepsilon^2 + \varepsilon'^2 + \varepsilon''^2 + \ldots .}{n-1}}$$

Probable error of a single observation is—

$$0{\cdot}6745\, e$$

Probable error of the mean result is—

$$\mathrm{E} = 0{\cdot}6745 \frac{e}{\sqrt{n}} = 0{\cdot}6745 \sqrt{\frac{\varepsilon^2 + \varepsilon'^2 + \varepsilon''^2 + \ldots .}{n(n-1)}}$$

Page 35, line 11 from bottom, *for* effected *read* affected.
Page 66, line 8 from top, *for* height *read* heights.
Page 72, line 11 from top, *for* E and F is, *read* E and F are.
Page 81, line 22 from top, *dele* and.
Page 87, line 11 from bottom, *for* of the ring *read* at the ring.
Page 90, line 19 from top, *for* Tyndall and others *read* Tyndall, Henry, Norton and others.

www.ingramcontent.com/pod-product-compliance
Lightning Source LLC
LaVergne TN
LVHW021419110826
845150LV00007B/1996

* 9 7 8 1 4 2 5 5 0 9 1 3 2 *